FAMOUS REGIMENTS

The Royal Dragoons
(1st Dragoons)

For a list of other titles in this series, please write to the publisher.

Other titles in this series by R. J. T. Hills:

The Life Guards
The Royal Horse Guards

FAMOUS REGIMENTS

Edited by
Lt-General Sir Brian Horrocks

The Royal Dragoons
(1st Dragoons)

by

R. J. T. Hills

Leo Cooper Ltd, London

*First published in Great Britain 1972
by Leo Cooper Ltd,
196 Shaftesbury Avenue,
London WC2H 8JL*

Copyright © 1972 by R. J. T. Hills

*Introduction Copyright © 1972
by Lt-General Sir Brian Horrocks*

ISBN 0 85052 120 3

*Printed in Great Britain by
Compton Press, Compton Chamberlayne, Salisbury, Wiltshire*

Acknowledgement

The writing of this book has only confirmed a lifelong conviction of the immense power of the regimental system which has brought the British Army through more than three centuries of wars and the perils of peace which can be even more dangerous. The Royal Dragoons are now part of a successfully amalgamated regiment, The Blues and Royals. But the co-operation I have received, not only from ex-Royals but from serving Blues and Royals of all ranks, surpasses all thanks that I can express in this small tribute to the oldest regiment of the Cavalry of the Line.

May I be forgiven if I leave out many names of those who have helped. Mention must first be made of the last Colonel of the Royal Dragoons and the first Deputy Colonel of The Blues and Royals, General Sir Desmond Fitzpatrick, who not only gave the undertaking his blessing but added very material help in advice and fact-giving. Further regimental assistance came from Major C. W. J. Lewis (now Career's Officer of the Household Cavalry). The score of the Regimental March came from Major E. W. Jeanes, Director of Music of the Blues and Royals. There was further encouragement from Mr 'Spike' Mays whose book *Fall out the Officers* gives a vivid picture of regimental life between the wars. My thanks are due both to him and his publishers, Messrs Eyre and Spottiswood for permission to quote.

The Royal Dragoons have been well served by their regimental historians. The ground work of this book was made possible by *The History of The Royal Dragoons* by Professor C. A. Atkinson, which brought the tale down to 1933 and *The Story of The Royal Dragoons, 1938–45* by J. A. Pitt-Rivers. Illustrations used are credited in the captions, except for those pictures which came from regimental sources. Those of the Second World War are taken from the official regimental account and were photographed by Major K. G. Balfour. More than ever I am indebted to Mr L. V. Shaw for patient help in searching out sources and pictures which could only be done in London.

R. J. T. Hills

Illustrations

Between Pages 10 and 11
1. The Tangier Horse capture a Moorish Standard
2. A Royal Dragoon, 1742
3. A Trooper of the Royal Dragoons after the Battle of Beaumont, 1794
4. 'The Charge of the Royals' at Waterloo
5. Cpl R. Wood of the 1st Royal Dragoons
6. Officer and Sergeant after Waterloo

Between Pages 42 and 43
7. Officer of the Royal Dragoons, 1832
8. The Royal Dragoons on Parade, 1843
9. Royal Dragoons turning out in 1848
10. The Duke of Cambridge inspects the Regiment
11. Officers of the Royals in Dublin, 1868
12. Officers at Riding School

		Facing Page
13	German Military Attaché at presentation of the Waterloo Wreath	50
14	Manoeuvres in the nineties'	50
15	1st Royals in Natal, 1899	51
16	The Royal Dragoons at Ludgershall	66
17	'B' Squadron at Vimy Ridge	66
18	On service in France, 1918	67
19	The regimental mascot greets the King and Queen, 1925	67

Between Pages 74 and 75
20. The Drum Horse of the Royals, 'Coronet'
21. The Dragoons drilling in Palestine, 1939
22. At the victory parade at Aleppo
23. A patrol relaxes in North Africa
24. 'C' Squadron in Normandy
25. The Regiment in Holland, 25 December, 1944
26. The Royals after the invasion of Germany
27. The advanced guard taking part in the liberation of Denmark

Between Pages 90 and 91
28. Officers at Shiban in the Middle East
29. A Security Check in Malaya
30. Training in Perak, Malaya
31. NCOs inspecting the obelisk at Southwark, 1959
32. The Royal Dragoons march through London, 22 October, 1963
33. 'The Last Dragoon'

The Royal Dragoons

INTRODUCTION BY
Lt-General Sir Brian Horrocks

The Germans had turned the Western Desert into a modern fortress in which the walls of olden times were replaced by deep mine fields covering their defences. Any breach which we succeeded in making was at once ringed round by enemy anti-tank weapons, and these narrow mine-free passages thus became a death trap for our armoured vehicles. Ten days after the start of the Battle of Alamein, we were beginning to feel that this well planned, well executed attack might ultimately founder in the mine fields. In November our last infantry reserves, the 151st and 152nd Infantry Brigades, backed by the 9th Armoured Brigade and two regiments of armoured cars were launched in a final effort called 'Supercharge', to smash our way out into the open desert beyond. 3 November, 1942, was a major turning point in the history of the Second World War.

From my Corps Headquarters on the southern flank, I listened anxiously to the roar of the guns supporting this offensive. At dawn on the Fourth I saw a figure running towards my caravan. It was Freddie de Butts, my Chief Intelligence Officer. Before reaching the door he shouted 'We're through! Our armoured cars have broken out into the open country beyond; both the Royals and the 4th South African Armoured Car Regiment are reporting by wireless that they are having a wonderful time, shooting up streams of Axis transport, heading for the west.' So victory at last was ours – but at a heavy cost. Walking over the battlefield afterwards, I counted no fewer than 85 tanks of the 9th Armoured Brigade in one stretch of a few hundred yards, all burnt out, but the Battle of Alamein was won, and the tide of war in the west turned for ever.

It was eminently suitable that the spearhead of this historic breakthrough should have been led by The Royal Dragoons, the oldest cavalry regiment of the line, who trace their origins back to the Tangier Horse, raised in 1661 by the Earl of Peterborough.

Good fortune and their role as heavy cavalry combined to protect the 'Royals', as they were always affectionately known and as they will for ever be remembered, from the rigours of Indian or Colonial service in the eighteenth and nineteenth centuries. Nevertheless they distinguished themselves in almost every European campaign. At home they fought at Sedgemoor, were with William at the Boyne and helped to suppress the '15. On the Continent they could normally be found wherever the fighting was, in Holland in 1702-3, in the Peninsula from 1706 to 1710, at Dettingen, at Fontenoy, back in the Peninsula under Wellesley and, of course, at Waterloo, where they captured the tricolour of the French 105th Infantry Regiment, the Eagle surmounting which eventually became their regimental cap-badge.

There is a tendency among cavalrymen to regard Waterloo as their own particular victory and Colonel Hills is no exception. He maintains that it was the charge of the two cavalry brigades under Lord Uxbridge which won the day. To my mind the charge is an excellent example of how cavalry should not be used and demonstrates the besetting sin of the British cavalry throughout their history – namely their lack of control. The officers were the very epitome of gallantry but they regarded each battle in much the same light as they would a day's hunting. The lesson which Prince Rupert's dashing horsemen were taught by Cromwell's less glamorous but highly efficient New Model Army remained unabsorbed by many of their mechanized successors in the North African Desert. Many a battle has been lost by gallant, yet senseless, charges against an enemy often dug-in and nearly always with superior firepower. The cavalry charge at Waterloo undoubtedly blunted the French attack and the Royals captured their Eagle, but it was not the turning-point of the battle. Indeed the British casualties sustained in the charge were out of all proportion to the advantage gained.

On the other hand, the Charge of the 'Heavy Brigade' at Balaclava on 25 October, 1854, was an example of a supremely successful and well controlled cavalry action. As the Russian Cavalry advanced on the 'thin red line' Scarlett launched his cavalry against the Russian flank. The Royals were in reserve. Suddenly their CO, Colonel Yorke, realized that their old friends, The Greys, were in trouble. Without any orders, the Royals seized the initiative, and his voice could be heard echoing over the battle, 'By God, The Greys are cut off. Gallop!' There broke from the ranks a cheer, as they advanced against the right flank and rear of the wheeling enemy mass. The Russians broke, for this time the Heavies were well in hand – particularly the Royals. In eight minutes it was all over, and the total losses suffered by the whole Brigade during this operation were barely 80. The Charge of the Heavy Brigade was one of the most brilliant operations ever carried out by British Cavalry, but owing to the glamour and publicity which has always surrounded the terrible Charge of the Light Brigade, where the casualties amounted to 300 out of 600 engaged, the splendid performance of the Heavies is usually relegated to the background.

The 14/18 War was not a Cavalry contest, and after the miseries and frustrations of those four terrible years, the Regiment had to be almost completely reformed. Yet that superb *ésprit de corps* and self-confidence returned rapidly. This emerges time and again in Colonel Hills' excellent book. Of this difficult period, between the two Great Wars, when the future of the Cavalry was very uncertain, he writes as follows:

> 'There is no such thing as a typical British Regiment, since the whole structure relies on its very diversity, but to visit the Royals in barracks between the wars was to sense the spirit that had endured for centuries, a loyalty to their own ideals. They took it for granted and had no need to parade what was obvious. All ranks knew where their duty lay, and if they were at times puzzled from above they gave no sign of it. If they knew that the horse was headed for oblivion they lived and worked and trained as if they would ride on in history. They read the blazons on their guidon with quiet, unassuming confidence, that if further honours were to be won they would be gained without fuss. It would be what they had joined up for.'

No-one could possibly express the spirit of a good Cavalry Regiment better.

It would seem as if in order to make up for the long periods of home service when they were classed as Heavies, the fates decided that no cavalry regiment should have a better record of service during the Second World War than the Royals; they were abroad from 1938 to 1950, and saw service in Palestine, Syria, North Africa, Italy and North-West Europe; and ended the war by being vociferously welcomed into Denmark. They were one of the last regiments to be horsed, but in 1940 they applied to be mechanized as an armoured car regiment, and were fortunate to be lent instructors by the 11th Hussars, veterans of armoured car operations in the desert. They took to this new form of warfare like ducks to water, and their grey berets now became a familiar sight at the sharp end of the battle, wherever it might be. I have already mentioned the historic role they played at Alamein.

I first got to know them during the breakout from the bridgehead in Normandy, when 30 Corps advanced from the Seine to Holland, covering 250 miles in six days against scattered German opposition. Our advance was preceded by a screen of superb armoured car regiments; from right to left the 2nd Household Cavalry Regiment, the Inns of Court, the 11th Hussars and The Royals. The latter also played a prominent part in that most difficult battle of all, Arnhem. On 2 May, the Regiment took 10,000 prisoners, and 7 May 1945 they ended the war which they had started in Palestine by crossing the Danish frontier.

The war was over, but there developed a constant demand for armoured cars in a peace keeping role, and the Royals subsequently saw service in the Canal Zone, in the Persian Gulf, in Malaysia and Singapore.

They have now, like many others, lost their individual identity, and on 31 March, 1969, at Detmold in Germany they joined the Household Cavalry, and became Royal by amalgamation with the Blues. It would be difficult to imagine a better combination than these two splendid regiments, the Blues and Royals, each with hundreds of years' sterling service to their credit.

Chapter I

Asked to epitomise The Royal Dragoons (1st Dragoons), one could fairly describe them as the very proud, oldest workaday regiment of the Cavalry of the British Line. They have, throughout three centuries of glowing history, done every service that has been demanded of them without advertisement, up and down and across the world. In the plain run of duty they have gathered their Battle Honours, from 'TANGIER, 1662–80', to 'NORTH-WEST EUROPE, 1944–45'. Every development in war has meant fresh effort. From 'Horse' they became 'Dragoons' and proved that the change meant little or nothing. They won their badge at Waterloo and charged with the 'Heavies' at Balaclava. They rode through the mud of Flanders and drew swords for the last time in battle in 1918. Almost the last to submit to mechanisation, they clambered into armoured cars in Egypt in the Second World War and headed west.

London-born in 1661, they started their service abroad against the Moors. When overtaken by amalgamation (that dread disease of the British Army) in 1969 they were abroad again. They had no homecoming and said no farewell to the place of their birth. They simply merged on a barrack square in Germany. They are now Household Cavalry, an integral part of the 'Blues and Royals'. But that is no part of the present story, which is simply that of The Royal Dragoons.

* * *

In the year 1661 soldiering was a trade definitely oversupplied. The nation that had welcomed back King Charles II had been governed for far too long by Oliver Cromwell and his major-generals. Old Parliamentary soldiers had been pelted from the London streets and the Treasury searched its coffers to pay them off and be rid of them. However Charles himself had other views: his father had lost his head because he had insufficient trained men to back him up whilst Parliament had triumphed by being able to call out the trained bands – especially those of London.

The King was determined that the new army (slightly camouflaged under the broad title of 'Guards and Garrisons') should be 'royal' and under his own control. The 'Guards' were not difficult to manage, since he was permitted to pay them himself, and only a few of the 'strong places' were garrisoned on anything better than a care and maintenance basis. Fortunately for his plans, he was able to build on the infant British Empire. Charles accepted, as part of his duty, that of getting married and a small, rather pathetic Portuguese princess was produced after some search. Part of the dowry of Catherine of Braganza were two outposts of the future empire – Bombay and Tangier.

If Tangier (key to the Mediterranean) were to be held it would need a British garrison. The claims of Portugal (and consequently now of Britain) were vague and hotly disputed. It was little more than a lightly fortified trading post which the Portuguese were glad to be rid of, but which was to be expanded under British rule into a bulwark protecting an important trade route and from which local rulers with a strong urge for piracy could be disciplined. The basis of the Tangier garrison was to be the infantry regiment which became eventually 'the Queen's', but there was required a small force of mounted men to keep the Moors well back from the perimeter. After its manner, the home government envisaged a little England beyond the seas: traders must be encouraged, a naval base set up with all its refinements and there must even be a mayor and corporation.

The new Tangier Horse sent its drums into London to raise the hundred troopers that composed its first establishment. They were not hard to come by: old soldiers, royalist or parliamentarian, roused themselves from their ale houses with enthusiasm. In just three weeks, on 21 October, 1661, there paraded in St George's Fields, Southwark, the troop of 'well-appointed horsemen' who in process of time became the Royal Dragoons. They had to wait three centuries before London claimed them, finally drumming their way back into the City – its only regular cavalry regiment – when in 1963 they claimed their new privilege of marching through the City with 'drums beating, swords unsheathed and Guidon unfurled'. Only a little earlier they had celebrated their tercentenary in Malaysia.

That they were 'Horse' was then considered a distinction, reserving for them a dignified, often decisive position on the battlefield. They were equipped with cuirasses, wore long scarlet coats over buff doublets, leather breeches and high boots which prevented them from fighting on foot anyway. Their arms were swords, a brace of pistols apiece and carbines: their horses were necessarily weight carriers.

As Tangier's first governor Charles appointed the Earl of Peterborough, not to be confused with the later earl who was to take the Royals under his command very much later in Spain. One of the swarm of needy cavaliers with a claim on the King's generosity, Peterborough had no cause for disappointment. He was governor, commander-in-chief, colonel both of horse and foot and captain – by deputy – of the troop of horse. As a cavalry colonel alone he had twelve shillings a day and ten as captain, with allowances for two horses. What did not please him was the meagre establishment of his horse: the more he and his successors saw of the place the harder did they press for an increase. The city had so extensive a hinterland that only cavalry could operate successfully against the swarms of expert Arab horsemen. The men from London had to fight even for their grazing rights.

Regimental descent is not always easy to maintain if it is to be achieved without a break in continuity but in the case of the Royal Dragoons there was never to be such a break. Even when Tangier was evacuated, it will be seen that its Horse was not disbanded. It came home as a unit and had become, even before it left Africa, the main ingredient of the regiment which became finally 'The Royal Dragoons (1st Dragoons)'. It should be noted here that this was the final official title of the Regiment. There have been various permutations but the Regiment has always been content to be known as The Royal Dragoons or more familiarly as just – 'The Royals'.

* * *

In January, 1662, a force of 3,000 men slipped down the Thames, the Tangier Horse occupying the *Tobias*, the *James* and the *Olive*

Branch, disembarking on 29 January after a voyage which was possibly the happiest of their experiences in this first trip abroad. The battle honour of 'TANGIER, 1662–80' (which was not granted until the present century) was not easily gained. It represents a life of constant harassment, endless sorties against a volatile enemy and all the health hazards then incidental to soldiering. The Moors soon took the measure of the scanty force, excelling in skirmishing, the use of ground and the setting of ambushes, while the British were slow to shake off European habits of warfare. The occupation of Tangier was one continual siege, loosely maintained, relieved only by the squabbles of the Moorish leaders among themselves.

Apart from battle casualties and the climate, the troops had their own habits of life against them. They were rough men living in a rough age, lacking even the few amenities they could look for in Europe. The city was populated by storekeepers, publicans, and the less reputable elements ready to prey on dissolute soldiery. The men were usually overworked and their pay hopelessly in arrears. But, as soldiers will everywhere, the men of Tangier built up some sort of environment for themselves, remembering their homeland by renaming features of the local landscape. They had their 'Whitehall', their 'York Castle', 'Whitby', 'Cambridge' and 'Kendal'.

Although by no means their first action, the Tangier Horse gained particular distinction in February 1664 under the Earl of Teviot, who had succeeded Peterborough. A fighter of resource, he was modern to the extent of using guard dogs to thicken up his sentry line. He constructed a line of forts which Ghailan, the current Moorish leader, recognised as an effective counter to his own plans of harassment and therefore sent in a full-scale two-day attack.

Main enemy effort was against Fort Charles, an attack pressed for four hours until Teviot decided to use his cavalry, now reinforced up to three troops, the junior commanded by Edward Witham. Teviot called him up and pointed out a red standard which seemed to be the chief Moorish rallying point. Witham needed no urging: sweeping down at the head of his troop, he cut down an enemy leader and brought off the standard in triumph. The superstitious Moors regarded the loss with so much dismay that 'They drew off in trouble, concluding the loss of their standard to be ominous, the like not having been done before'.

Teviot, whom the Moors took for the Devil himself, did not live long to enjoy his reputation, but it was extended to his horsemen, who pushed their outposts ever outward. Early in May, with Ghailan burning for vengeance and bringing up every available man, Teviot met him on a sortie at the head of his cavalry and seven columns of foot. They smoked out several minor ambushes and became so over-confident that they started to graze their horses, which experience should have taught them was too great a risk. Ghailan swooped and, though most of the mounted men got clear away, the foot, hopelessly outnumbered and surrounded, their ammunition failing, finally stood to be slaughtered, losing Teviot himself, 19 officers and over 400 men. But British hold on the town had been consolidated and Teviot is described as 'the first soldier of distinction under whom the Royals had the honour to serve: the first in a long line of men who have led British regiments to victory against a formidable African enemy'.

* * *

Although it was at one time proposed to disband the Tangier Horse altogether, better counsels prevailed and the cavalry, far from being reduced, were brought up to something like the strength of a regiment. But at one time it became so low that the effective fighting strength was down to ten men, the senior being a farrier – without a horse! Attempts were even made to recruit Spaniards while infantry officers were encouraged to keep horses on which a few men could be mounted in emergency. In 1680 the garrison included, besides its original units, such notable fighting corps as the Royal Scots and a composite regiment with a strong Foot Guard's element. Six new troops were ordered to be added to the Tangier Horse, of which three actually reached Africa. Again it was easy to raise men and even the horses were 'old soldiers', the new troops being allowed to call upon both Life Guards and Blues for remounts.

The job of yet another new governor was formidable. Sir Palmes Fairborne found his garrison in rags and had twice to suppress mutiny. But with his reinforcements he was able, in September 1680, to go over to the offensive, the most pressing need being to recover Pole Fort, a key position now in enemy hands. Fairborne himself fell in a preliminary engagement, but on 27 October, 1680 was fought the final serious action of the Tangier garrison, in which the Moors were routed and agreed to a six-months' truce. It was this fight that limited the dating of the battle honour, the territory being at last deemed a 'peace station' although the troops were apt to wonder why!

Tangier was in any case an impossible outpost under the conditions of the seventeenth century. 'For twenty-one years', writes one historian, 'Tangier had proved a constant strain on the royal finances, seldom absorbing less than £70,000 a year. A gold mine to many of his subjects, Tangier had come to present itself to the King as a hole in his pocket'. Thus in 1684 Lord Dartmouth was sent out to supervise evacuation and destruction of the fortifications. The proceedings were duly reported by Samuel Pepys, who went out as Dartmouth's secretary. Everything the precise little official saw attested the impracticability of the place. The walls were everywhere overlooked, with the very water supply dependent upon the whim of the surrounding Moors. 'It is a place of the world', he summed it up, 'I would last send a young man to, but to Hell'.

Both the place and its governor were anathema to the meticulous civil servant from London. All he could see in Kirke, adored though he was by his soldiers, was 'an overbearing redcoat, who outraged every decency and mocked every prudent, diligent rule that his own (Pepys') experience had taught him to honour'. But then Kirke, who was rather more than a plain product of his times, admitted: 'I don't pretend to be a saint'. Pepys was so horrified by it all – 'vice in the whole place for cursing, drinking and whoring' – that he forgot his prudent custom of using shorthand and spelled it all out in plain, trenchant English. It was, he said 'a military Bedlam'. Fortunately the withdrawal did not mean the end of the regiments which formed its garrison.

Chapter 2

THOUGH forced to draw in his horns abroad, Charles II was by this time sure enough of himself to save the units from Tangier from extinction. The infantry came home to become the 2nd and 4th of Foot, the former remembering down the centuries their forthright colonel in their nickname of 'Kirke's Lambs'. The horse was expanded in slightly different guise. They were exactly what the King needed – men tried in his own service, professionals loyal to the growing regimental idea that was to become the kingpin of the British army system. As early as November 1683 commissions had been issued for the embodiment of a regiment to be called 'Our Owne Royall Regiment of Dragoones', while included in Dartmouth's orders for the evacuation of Tangier were instructions that the four troops of horse must be brought home intact to join two other troops now raising in England. With an unbroken record dating from 1661 the newly-titled Regiment could lay undoubted claim to be the oldest regiment of British line cavalry.

The 'dragoon' needs explaining. He was a continental invention, a mounted infantryman whose pattern of musket was called, by the French, a 'dragon'. The men were skirmishers who could advance rapidly to a battle station, often used to line hedges over which an enemy could advance. They were first seen in England during the Civil War. Rupert used them effectively at Edgehill and they saved Oxford by holding the vital bridge at Cropredy in 1644. Cromwell raised a small force for the New Model Army but the new 'Royall Regiment' was the first on the Restoration establishment.

There was an economy that appealed to the Treasury, since dragoons were paid less than 'horse', were less expensively equipped and had to be content with 'good squat dragoon horses' instead of chargers. What rankled in the minds of the men from Tangier was that they now had to give pride of place to regiments of horse raised after their own corps and who, in course of time, became dragoon guards. But the title 'Royal' which came to them as the senior regiment of their arm was some consolation and at all events an improvement on the common practice of being named after their colonel for the time being. There are certainly occasions when the Royals themselves are thus referred to but these are rare and 'The Royals' have normally been – just that.

To meet the niceties of military custom the corporals of horse now became sergeants and, instead of trumpeters the dragoons had drummers and hautboys. But to put a soldier on horseback is likely to give him a better conceit of himself and by the time they fought at Sedgemoor in 1685, there was little, except perhaps their mobility, to distinguish between dragoons and cavalry proper. The whole plan turned out to be a non-workable compromise and one critic wrote: 'When mounted they are taught that no infantry can withstand the impetuosity of their charges: when drilling on foot they are taught to consider themselves invulnerable against cavalry'.

* * *

The transformation brought them their most noted colonel in the person of John Churchill, with Lord Cornbury, son of the 2nd Earl of Clarendon, as Lieutenant-Colonel and effectively commanding officer. Churchill's appointment called for some comment, since the future victor of Blenheim, despite some good fighting success, was regarded mainly as a man about court. The wits had their own sally:

> Let's cut our meat with spoons!
> The sense is as good
> As that Churchill should
> Be put to command the Dragoons.

In any case the Tangier Horse became dragoons on 1 May, 1684 and were joined by their additional troops, regaling the 'young 'uns' doubtless with fearsome stories of fighting against ferocious Moors. But the Regiment was warned that what passed for normal conduct in Africa would hardly be appreciated at St Albans and Hertford, where it was first stationed, moving later into Berkshire. The Royals came 'home' to Southwark for a few days in September, to prepare for a royal review (King Charles II's last) of the whole army on Putney Heath. The royal status of the Regiment was emphasized, troop standards each bearing a badge of the Royal family. But these were no parade troops: with no war to provide real active service they had enough internal security work to keep them busy, escorting the mails, making roads safe for travellers and assisting the coastguard. This last was the most popular as the men were given extra pay and sometimes a bonus on confiscated goods.

* * *

That James II ever succeeded to the throne was a misfortune that took many by surprise, he being considered a much worse 'life' than his normally robust brother. It was an advantage that he had two Protestant daughters to succeed him and that his second, Catholic wife had a tendency to miscarriages. He therefore succeeded to the throne in a general atmosphere of loyalty which he knew he owed largely to his deceased brother and therefore did not trust. There was a drawing-in of forces to the neighbourhood of London, the Royals being first quartered at Hounslow, Brentford and Uxbridge being put under canvas later in Hyde Park.

James had not long to wait for trouble, which came first from the north, where the Duke of Argyll landed at Cambelltown to raise rebellion. The Royals were given a swift route for Carlisle, two troops covering 290 miles in 20 days. The rest of the Regiment did not march, fortunately as it turned out. Argyll was easily disposed of but in the meantime the Duke of Monmouth, Charles II's eldest illegitimate son, had landed at Lyme Regis, with far more favourable chances. He was the 'Protestant Duke' and he had the largely Puritan west country to draw upon for recruits.

King James showed in his best light, with prompt and efficient orders for a concentration of troops. Two troops of the Royals marched in advance for Salisbury, the remainder being drawn in to London, while the Regiment was augmented by five new troops. The general concentration under Churchill and the march of the main army under Feversham (the commander-in-chief) ended inevitably on Sedgemoor, where the Royals played their part. When the rewards were handed out they brought the colonelcy of the Royal Dragoons to Cornbury, Churchill taking over the Third Troop of Horse Guards, whilst the Regiment had the dubious honour of providing a detachment on Tower Hill when the rebel Duke was executed.

* * *

The 'Bloodless Revolution' of 1688 can only be touched on here insofar as it affects our story. James seized on Monmouth's attempt as an excuse for a general expansion of the army which brought the Royals their share of promotions and even brought commissions to senior N.C.Os. and quartermasters. The latter were warrant officers who only existed in the cavalry and were the forerunners of sergeant majors and quartermaster sergeants. They purchased their warrants, usually from the ranks, but were often young men of good family on their way up by a somewhat cheaper ladder than the direct purchase of an officer's commission. James felt that he could rely on the army he had so largely built and trained, for as he so justly if sadly claimed: 'Never any prince took more care of his sea and land men as I have done and been so very ill repaid by them'. He discovered too late that the troops merely reflected the mood of the nation.

History, as so often, turned on a comparative incident. Mary of Modena, James' queen, gave birth to a Prince of Wales who would obviously be brought up a Catholic. Plotters had for long been in touch with William of Orange, husband of James' elder daughter. They now sent him an invitation to come over. There was an agonising wait as men studied the weather vane – which once dictated a false start – but William landed duly at Brixham and the weapon that James had forged with so much care snapped in his hand. The King felt confident at Salisbury, with a trained army of 40,000 to oppose William's 14,000. The Royal Dragoons joined on 10 November but Cornbury, their colonel, was deep in the conspiracy. He, with Compton of the Blues and Langston of the Princess Anne's Regiment, marched their regiments under forged orders to Axminster, determined to hand them over intact to William.

1 *An incident in the Moorish Wars – the Tangier Horse capture a Moorish Standard, 1664*

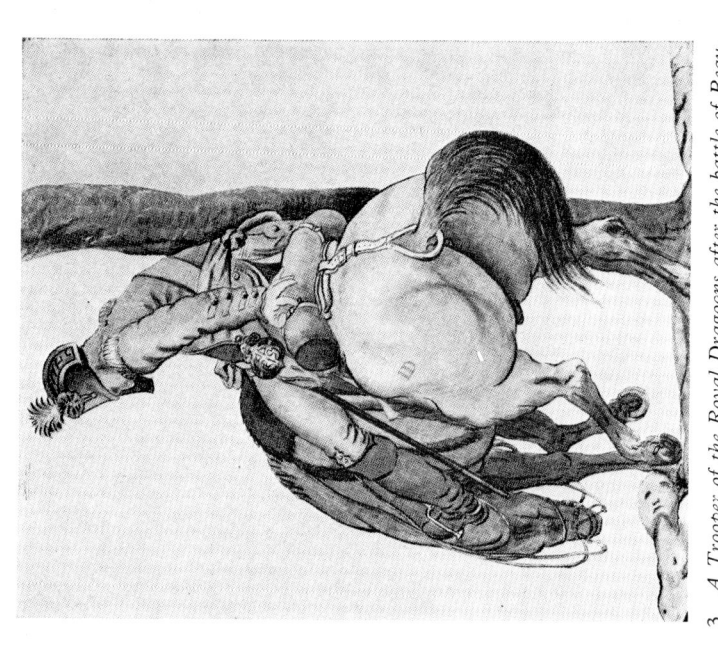

3 *A Trooper of the Royal Dragoons after the battle of Beaumont, 1794*

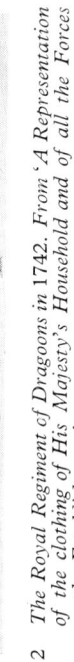

2 *The Royal Regiment of Dragoons in 1742. From 'A Representation of the clothing of His Majesty's Household and of all the Forces on the Establishment.'*

4 'The Charge of the Royals' at Waterloo.

5 A veteran of the war in Spain, Corporal R. Wood of the 1st Royal Dragoons, from a series of vivid sketches by Thomas Heaphy made in 1816.

6 Officer and Sergeant of the Regiment in the period immediately following Waterloo

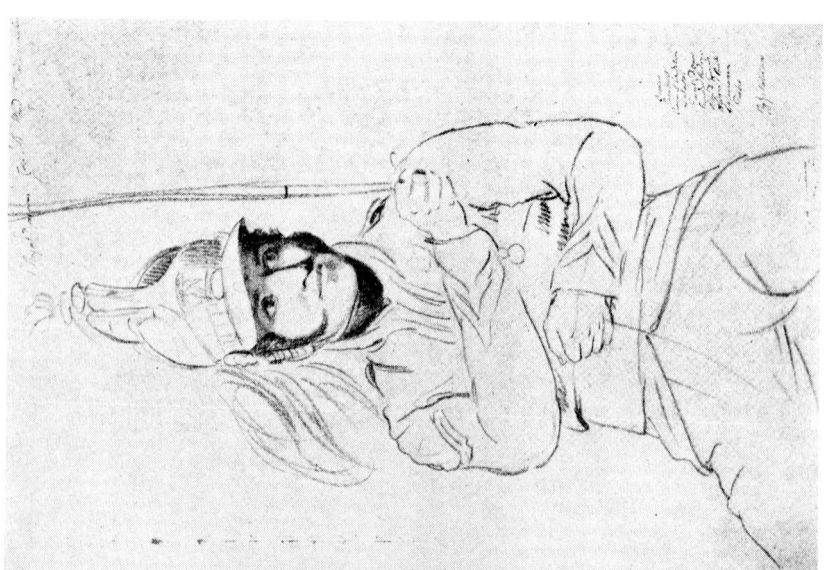

Loyalists among them now became suspicious and Major Clifford of the Royals insisted on seeing the supposed orders. The plot collapsed, most of the junior officers and men marching back to rejoin their monarch. But it was the beginning of the end brought about mainly by the King's hesitation. It is not a pretty story but 'if ever a man deserved to be deserted, James did'. With scarcely a shot fired in his favour, physically ill, hesitant where only boldness could have served, he turned his back and, in his flight, did his best to leave the kingdom in a state of turmoil. Pocketing the handiest of the crown jewels, he flung the Great Seal into the Thames, ordered the army to be disbanded and fled to France. William cared little for the crown jewels and less for the Great Seal but it was the British Army he had really come to take over, to join in his lifelong war against France and the Catholic religion.

Orders were at once issued for its re-assembly and the Royal Dragoons came in to Farnham and Alton. Everyone likely to remain loyal to James or his religion was purged, to the extent that of 25 officers serving in the Royals at the start of 1688 only eight remained by the end of the year. That the Regiment survived was in the main due to the hold that the regimental system had gained. What happened outside cantonments was beyond the consideration of the ordinary soldier. His loyalty was to the Regiment – the club – with its traditions and standards.

There was little love lost between William of Orange and his reluctant subjects. He had landed at the head of 14,000 rapacious mercenaries of assorted nationalities. Thirty per cent of them were Roman Catholics, for whose discomfiture they had ostensibly come. It was the Dutch who were the traditional enemies – not the French – and London fumed at the sight of their dingy blue uniforms mounting the Guard in Whitehall. But William had no doubt of the fighting quality of his British troops.

First he must establish dominion at home. There was already rebellion in Scotland and the Royal Dragoons were among the troops who headed north at a hot pace, suffering more from marching over difficult country than from action. In October of 1689 they crossed to Ireland, 'weak and worn-out' as William's General Schomberg described them. They fought under their new king's command at the Boyne, although there is little information as to regimental action. They were shipped back to England to meet a threat of invasion, only to return to Ireland for the winter of 1690.

* * *

This time there was more real campaigning, with William besieging Limerick for the second time. The perimeter of the city was too large for the available infantry and the dragoons filled in the gap. Siege duties apart, the Royal Dragoons were never idle, since the scene was enlivened by freebooting gangs of *Raparees* fighting as much for their own ends as for those of James II. Near Castleford a detachment of the Regiment killed over 40 of these troublesome gentry, while on another occasion Quartermaster Topham, commanding dragoons and infantry on country ponies, killed seventeen and took nine, hanging seven on the spot. The patience of British soldiers is not endless and the Royals were angered by the frequent murder of comrades.

Back at Limerick it was the Royals who spear-headed the vital attack which completed the investment. It all ended in October 1692, when a capitulation gave the garrison full honours of war, a great portion taking service in the famous Irish Brigade of France. The Royal Dragoons were now free to return home for their continual round of police duties – plus even a spell for some of them aboard the Fleet. Such duties kept them away from the battles of Steinkirk and Landen in Flanders which, if they were defeats for William, added much to the growing reputation of British regiments.

The Royals thought their turn had come at last when they crossed the North Sea in May of 1694, but there was no set-piece battle with a job for mounted men. That the Regiment had been active in skirmishing – probably back at their trade of dragooning – may be seen from the fact that, when winter set in they had to ransom eighteen prisoners, the highest of the sixteen cavalry regiments in the field. The following year proved equally disappointing and the Peace of Ryswick in September 1697 brought the Regiment home. Parliament was clamouring for a general reduction of the army and above all for the Dutch troops to be sent home. The Royal Dragoons were reduced to 283 of all ranks and they resumed their dreary round of home duties. The troops were so scattered that the commanding officer complained that it took him three weeks of travel even to visit his regiment.

William of Orange, ailing, unpopular but indomitable, did much to establish the British Army, leaving it to his successor in a fit state to embark upon some of its most glorious campaigns. But it should be remembered that it was a Stuart creation. Charles II planned and James II forged a weapon capable, if only it had been commanded and used, of chasing the invader into the sea. King James II had not the temperament to lead as a general but his annual training camps shook his troops down into an army. The Royals had seen service in North Africa, at home, in Ireland and in Flanders – but Hounslow Heath played a greater part than they even realised at the time.

Chapter 3

WILLIAM III's legacy to Britain was the long war against Louis XIV in which British soldiers gained glory; their greatest general immortality; and the politicians showed their talent for throwing away what their armies had won. The war, to which the Peace of Ryswick had merely dotted in a semi-colon, now became the War of the Spanish Succession, revolving around Louis' determination to set his grandson Philip on the Spanish throne, despite his own promises and the rival claims of the Austrian Archduke Charles. The rôle for which the Royal Dragoons were cast was a tricky one. They paid a brief visit to the Netherlands, had a first view of the Iberian Peninsula and ended up as prisoners of war!

* * *

British decisions on the international level were still insular and even parochial. It was the death of James II and the recognition of the Prince of Wales by Louis XIV that enraged Britain to the point of war. William of Orange himself died in March 1702, but his troops were already mobilising and he had made sure that Marlborough was available to take the bâton from his hand. At regimental level the Royal Dragoons were on their home ground, at Southwark and Rotherhithe waiting to embark. Their colonel was now Lord Raby who, despite a diplomatic post in Berlin, exercised a surprisingly close control of his Regiment during the next few years. Most regimental colonels took a great pride in their regiments, were intensely jealous of their rights but expected to make a profit on the job. Their standing enemy was the Treasury, whose form of book-keeping in the age of Queen Anne was complicated in the extreme, giving the civil service a lead over the nation it has never surrendered.

Augmentation to war establishment was easy, since William's forced reductions had put thousands of fighting men on the streets:

but the vacant commissions caused some headaches. Two retired captains were brought back but Raby drew the line at Lt Cropley, who had 'married an innkeeper and notoriously quitted the service to do so', while Cornet Marshall had sold out so abruptly that he was regarded as a quitter. Ex-Quartermaster Richardson was suspect as to loyalty, sheer inefficiency and for having married a woman of even lower status than an innkeeper! The practice, long to be continued, of giving commissions to infants, was seen a little later when William Wentworth was given a cornetcy and indefinite leave – 'he being but a child' – but he served for 44 years.

In the rickety small ships normally assigned as troop transports it was difficult to get to the continent at all. Bad weather in the Thames held up and eventually cancelled the Royals' embarkation scheme. One troop leader had 'never seen such a ship' as his. She ran aground, fouled other vessels and had to be beached with two feet of water in the hold, when the horses were taken off. Finally the Regiment crossed the Thames in despair and made for Harwich, arriving on the other side in such a state that the men were put into quarters and the horses grazed for a month. By August two squadrons stood complete and fit to campaign. But for two seasons Marlborough was forced back on to the old-style campaign and the cavalry could only seek minor distinction in skirmishing round besieged fortresses.

Then came the ray of hope which affected the Royals closely: a second front. Portugal remembered her old alliance with England and, by providing troops and a firm base, made it possible to send in the Archduke Charles to fight for his crown where his presence might be most effective. From his own force Marlborough was instructed to detach one regiment of dragoons and four battalions for the new expedition. He was broad-minded enough to select first-rate troops and the Royal Dragoons could therefore take it as a compliment that they were chosen.

There were complications enough in the transfer. The Treasury seized on the fact that the Regiment was to sail dismounted and withdrew the forage allowance before even a horse had been disposed of. The Royals won this paper battle and handed over their horses at a reasonable price to other regiments in the theatre. Ordnance was deeply shocked when the Regiment insisted its arms were not fit to fight with. Indeed the men had come to prefer swords that buckled in their hands rather than pistols which exploded in their faces. Taking ship, even without horses, was again a deadly peril. Although first warned in September 1703, it was in the middle of the following March before they saw the Tagus.

The voyage was the usual hazard of the times – not the least of the dangers to which Queen Anne's soldiers (and those of her successors for 150 years) were liable to be exposed. These expensively recruited and trained men were normally committed to the deep in tubs barely capable of sailing down the Thames, with crews far below standard. Except for providing convoy the Royal Navy ignored such lowly duties as transporting sea-sick redcoats.

'It was a universal complaint', wrote one diarist, 'the description of boats that were hired as transports. Not only were they wholly unfit for the service but in many cases they were unsafe as not being seaworthy. It is only necessary to look back to see the number of brave fellows who were lost as being exposed to those old, crazy, unserviceable boats'. A private of the Guards was more picturesque when he described: 'continual destruction in the foretop, the pox above board and the devil at the helm'.

* * *

The Regiment sailed under the command of Lt-Col Robert Killigrew but perhaps its most notable figure was that of Lt-Col James St Pierre, who eventually took command in 1716. He came from one of the refugee Huguenot families to many members of which fighting was a natural profession. St Pierre had received his cornetcy under Cornbury and by now was often far higher in allied councils than his actual rank seemed to authorise. In addition he left a narrative which is an important historical source. The Royals had – rare for those days – another (this time anonymous) writer from the ranks whose story, with all its semi-illiteracy, has come down to us. In his own words he had 'entertained himself as a soldier in Her Majesty's Royal Regiment of Dragoons' on 26 February, 1702 and joined them in the Low Countries.

In Lisbon the Royals discovered the wide gap between promise and performance normal in the Peninsula. The King of Portugal had agreed to mount the Regiment on arrival and had indeed collected some animals with four legs. These had been considerably reduced for want of care and forage, half-starved and 'scabby' – not at all up to the standard a proud regiment demanded. Moreover the cavalry brigadier, Harvey, had his own regiment, the Bays, with him and by the time he had seen to their needs, there remained only 170 sorry screws to divide between the Royals and the Walloon Dragoons. The Royals were particularly enraged in that Harvey ignored the ruling of King Charles II which gave 'post to ye Dragoons of all those raised after them, especially our own Regiment'. Had his ruling been adhered to it would have changed the whole order of precedence in the Cavalry of the Line.

There was little prospect of immediate action, the Portuguese forces being split into small detachments merely capable of speculating on the possible action of the enemy. The French were commanded by the Duke of Berwick, Marshal of France and son of James II by Arabella Churchill and therefore a nephew of Churchill. Command was an additional difficulty, common in allied armies of the period. It was divided between the Earl of Galway (another Huguenot) the Dutchman Fagel and the Portuguese das Minas.

Lack of real action did not prevent casualties from weather and unhealthy stations, so that sixty men of the Royals died that year and Raby learned in a letter: 'Pray God you may never see this hellish country, all are sick of it and everyone ill. Many say it is more unhealthy than the West Indies. They starve our horses first and then us'. When St Pierre mustered the Regiment in September he had only 296 privates fit for duty, with 47 sick. He admitted that he had never seen the Regiment in worse condition.

* * *

Next year brought on the scene a name familiar to the Royals, a relative of their first colonel, Charles Mordaunt, 3rd Earl of Peterborough, with vague orders, a new expeditionary force and a double commission as commander-in-chief both at sea and on land. Peterborough was a brilliant but erratic general, used to mixing with the great, garrulous and plain-speaking to the limit of courtesy. On one occasion during a tour of inspection in England he reported of one newly-raised regiment: 'It wants nothing to complete but clothes, boots, arms, horses and men'. Obviously not apt to be a general favourite, he was later summed up by Swift as: 'The ramblingest lying rogue on earth: the old hangdog I love him dearly'.

Peterborough reached Lisbon on 20 June, 1705 with a strong force of mainly raw troops and loosely-worded instructions to open up a new front in the Mediterranean, subject to a council of war which included the Archduke, Prince George of Hesse-Darmstadt, Lord Galway and the Portuguese commander. Of these Galway was the most helpful, a practical soldier who gave Peterborough two regiments of dragoons – the Royals and Pepper's – and authorised him to take two seasoned regiments from Gibraltar in place of some of his new battalions. Gibraltar had been the outstanding success of the war to date and in its capture Prince George had played a distinguished part.

Savoy or Italy (recommended in Peterborough's London briefing) were deemed impracticable for the time being and it was decided to make a descent on the east coast of Spain itself, where there was said to be enthusiasm for the Archduke, who was to sail with the expedition. The Royals were reasonably fit for the adventure, up to strength and mounted, though some of the remounts had not yet been shod. Sailing quietly round Spain, the expedition first put in at Aldea Bay to water – and to proclaim Charles King of Spain and the Indies.

Peterborough was aflame for a quick dash at Madrid itself but was persuaded instead to go for Barcelona, a fortress with a garrison larger than the investing force. There was of course the Fleet, though Peterborough was apt to confuse matters by issuing one set of orders on board ship as admiral and then countermanding them when he became a general on dry land. He was careful with his cavalry, as he had to be, the ground being boggy and the enemy having burned up all the forage they could reach. The Spanish irregulars, or Miquelets, failed to turn out to fight for their liberation in anything like the strength promised.

The Navy landed guns and Barcelona was duly invested in what appeared to be such a barren enterprise that orders had actually been given to re-embark when Peterborough produced a new plan. This, said to have been thought up by St Pierre, involved a surprise attack on the isolated Fort of Montjuich, less than a mile from the city walls and overlooking them. Peterborough and Prince George had not been on speaking terms for days but the latter was won over by the glib Englishman to take part in the attack, which was made at the end of a night march, covered by the Royals.

'Our regiment of dragoons', says the nameless diarist, 'was ordered to dismount in order to follow the attack but we did not, for we were countermanded for our foot prospered in their attack and was soon masters of the castle'. As a consequence Barcelona itself surrendered, much to the relief of the allies, 'for there come such heavy raines, enough to drowne us'. They marched into the city amid general rejoicings. Prince George had fallen at Montjuich but Peterborough took his share of the spoil in the person of the Duchess of Popoli, reputed the most beautiful woman in Spain, with a husband conveniently absent.

Charles, 'king' by proclamation, was content to rest on his laurels but Peterborough sought fresh worlds to conquer, finding a suitable one in Valencia, where the population was friendly and horses and supplies plentiful. His forces hardly matched his bluff — of which he had ample — but he made skilful use of his cavalry, sending them ahead in small packets, so disposed that they were taken for heads of non-existent advancing columns. He gained the reputation of 'taking walled towns with dragoons.' The men themselves reacted to such an adventurous commander. One of his better efforts was to ride out to meet Barrymore's Foot at the end of a weary march. Would they not, he asked, prefer to ride? Almost too breathless to reply, the men stumbled after him up a hill, on the other side of which waited six hundred horses, with his lordship's secretary, pen in hand and prepared to give the officers of Barrymore's new cavalry commissions into Pearce's Dragoons.

The Royals marched via Tarragona to Tortosa and thence to San Mateo to drive off the Irishman Mahoney with the loss of his baggage. They were less lucky at Peniscola where, to their surprise 'the Governor did not care a pin for us or what we could do, so we marched back to Bennicarlo'. The recompense came at Murviedo, a hostile city where 'their pigs we did feed on and their wine we well drunk and that without being measured to us'. War sometimes supported war in those days. Valencia itself was a triumph, even the friars taking up arms for King Charles.

But one dragoon of the Royals met his match in an 'Irish Dragoon (from the other side) that had lost his horse. His arms he did give me but said he had but one piece of silver as I would not take that'. But the men of the guard under whom the wretch spent the night were more thorough, for they found he had above 30 gold pieces on him! There were plenty of Irishmen on the other side, Mahoney himself being a relative of Lord Peterborough, with Lady Penelope O'Brien a common aunt.

* * *

Madrid, though less of a capital than most in Europe, could not fail to attract and was in fact occupied by Galway, starting from the opposite coast. Charles himself had to be spurred to the effort and Peterborough, disgruntled by overall command having been given to Galway, came to Madrid direct from Valencia instead of riding in Charles' train from Barcelona. The Royals were by now hardly a parade regiment. Weather and bad supply arrangements had so plagued them that they rode into the capital short of 80 men and 150 horses, half the latter being shoeless. Their cloaks had been cut down to make jackets and the men, from lack of proper administration, had turned to plunder to feed themselves. Madrid did not welcome its deliverers – an army largely Portuguese led by a heretic.

It was the tattered state of the Regiment that saved it from the disaster of Almanza, 'we being a long time in the country and was bare of clothes . . . our horses but weak, so we were ordered out of the camp before the engagement'. Thus they found themselves back in Valencia. To sketch further operations in detail, however arduous they were for those engaged, would be tedious but two major engagements gave the Royal Dragoons more chance for distinction. Peterborough, in a rare huff from being superseded a second time, took himself off for a grand tour of Europe, leaving Stahremburg in command, with Stanhope at the head of the British contingent.

Madrid was again the objective of an allied campaign and the battle of Almenara, fought on 16 July, 1710, came near to deciding the war in Charles' favour. That victory was not complete was due to Stahremburg's caution, in refusing to let Stanhope and his British loose until just before sunset. To be subject to Stahremburg's caution was a piece of ill-luck since an earlier proposal to send the famous Prince Eugene had been overruled. As it was Almenara was a triumph for the British cavalry. Stanhope, seizing a rare opportunity, formed a complete line which swept the Bourbon troops before it, despite their numerical superiority. The Royals were particularly prominent in the pursuit, under a new commanding officer – Edward Montague whose job had cost him £4,000. 'Had we had but two hours more daylight', laments the narrator, 'we had with the blessing of God undoubtedly gained the crown of Spain that night'.

There was further victory at Saragossa on 9 August, which the Royals helped to win – on half a bread ration and some water. The Regiment charged against a squadron of heavy horse and the diarist was lucky in that: 'I had not one drop of blood drawn from me that day but my hat was cut all to bits and my coat was shot through in many places, with a balle through my waistbelt'. Madrid stood open and this time, at the start of three days' rejoicing, welcomed the Allies with 'great tubs of wine in the streets'. But it still proved untenable and the Allies were now opposed by Marshal Vendôme in his most dangerous mood. There was an effort to get into winter quarters in Aragon but the enemy gave them little rest and now, as the writer puts it, 'begins our tradage'.

Stanhope marched for Brihuega, a black name for the army. On the march the two regiments of dragoons were most ably handled by Pepper their brigadier who, having been fairly caught napping at one point announced: 'Gentlemen, I brought you on and I will carry you off', commencing an orderly retirement in which he himself played the part of a squadron leader. Thus, on Saturday 25 November, 1710 they came to Brihuega, a weak place with old Moorish walls, incapable of defence. Stanhope, relying on support from Stahremburg which never came, took few precautions until Vendôme, well-informed of his plight, came up with him.

Stanhope's men made a noble effort but the enemy broke into the town through an inn built up against the crumbling wall and 'knockt us downe as fast as they could shoot'. Finally, with the enemy within a dozen yards of a last breastwork, orders were given for 'our Drum Major to sound a parley'. Thus twelve British regiments, including the Royal Dragoons, were marched off prisoners of war, treated abominably, prodded along on foot by their mounted escort, stripped when they fell, plundered of everything, spending a Christmas in Burgos on bread and water. Some allowed themselves to be recruited into the enemy ranks and then took the first opportunity of deserting back to their own side. Two entirely new regiments were recruited in Catalonia by Major Lepell of the Royals merely from this source.

Rumours of peace were rife in the spring but the remnants of the Royals had to wait until September before they were marched to Bordeaux, only to find that they were in the wrong port. Eventually, in the *William and Mary* from Passages they came into Sandwich and were posted to Doncaster. Their luck was in at last, for they were the only returning cavalry regiment not sent to the lower-paid Irish establishment.

Chapter 4

THE War of the Austrian Succession and the Seven Years' War were really one. Whatever the official causes of dispute, the real British aim was the destruction of French predominance in Europe and the reduction of her growing colonial power. If any other grounds were necessary they were to be found in the fact that Britain's rulers had a hereditary interest in Hanover, which must be protected, however much it might itself contribute in the shape of troops – and this was sometimes considerable. Britain was confident in her naval power and had rarely to complain of her soldiers when they were committed in Europe. As mid-century approached the army was solidifying into a formidable force, though its strength was largely in the regimental system which suited the islanders best. The stiff little German princes who had come in in 1714 may not have been popular but their handling of the army they took over was usually efficient. A series of training camps was inaugurated in 1739 in which the hand of George II was evident. There were camps at Windsor (where the Royals took part) and elsewhere, repeated a year later at Newbury and Devizes. Commanding officers rejoiced at the chance of getting their regiments together for training. England was already at war with Spain but this was purely a naval affair and Britain had not taken the Spanish navy seriously since 1588.

Weaknesses there were bound to be in an army which had not fought for so long. These were accentuated by the system which forced wealthy young men ahead and retarded the aged poor who could not purchase their steps. When at last the Royal Dragoons were warned for foreign service they left behind two veteran lieutenants. One with 30 years' service in the rank had a 'palsy' and could not mount his horse, whilst the other had a 'rheumatism'.

* * *

Early in 1742 King George II decided on active intervention in the cause of the Empress Maria Theresa of Austria, acting under a pledge to support her accession given by him as Elector of Hanover. She had already lost Silesia to Frederick the Great of Prussia and was now under attack by France and Bavaria. The King was not anxious to cross swords with Frederick, whose territories marched with his own and he had in addition a strange quirk which decided that he was not at 'war' with France and only acting as an auxiliary to Maria Theresa. The Dutch, to add to his troubles, were not at all keen on the war anyway, although they formed part of the alliance. An allied army was formed in the Netherlands under the veteran Lord Stair who, together with the Monarch himself, had won his spurs under Marlborough. There were 16,000 British troops with 6,000 Hessians and 16,000 Hanoverians, all paid by Britain.

Lack of green forage kept the troops back for most of the year and George, athirst for military glory, had a careful eye for Hanover. There was therefore the prototype of a 'phoney' war until the campaigning season of 1743, by which time the Empress had driven out French and Bavarians and was able to concert a plan for attacking them during their withdrawal. Lord Stair moved up the Rhine, crossing at Andernach. The troops were welcomed for their unusual habit of paying for what they consumed, appreciating the beer but despising the wine as 'washy stuff'. The supply problem forced Stair further on to the Main, where he was offered a fair chance of success by an opponent who was not the most brilliant example of a French commander-in-chief. He was restrained by the King who now came up to the army.

De Noailles now set his 'mousetrap' which resulted, against his careful calculations, in the allied victory of Dettingen and the last appearance of a king of England in battle at the head of his troops. The allies were desperately hungry, with the enemy ahead, behind and to one flank, the other being taken care of by a line of wooded hills. Early in the morning of 16 June the allies started to move on Hanau under fire from across the river. The little monarch was happy enough: Stair less so. The latter discovering a force under de Grammont, de Noailles' subordinate, on the move to the allied right, 'thought it my duty to meddle' and moved cavalry to the threatened flank.

In the new line-up the Royal Dragoons held the extreme right and escaped the preliminary bombardment. It was now de Grammont who was the mouse to snap at the bait. He rushed in to speed up what he saw as an allied retirement. Horse and foot failed to co-ordinate and gave the Royals – among others – their chance. The French Black Musketeers, shaken by British platoon fire, swerved to avoid it and were confounded by Stair's cavalry. The Royals dashed in with complete success, an unnamed sergeant taking a standard. It was described as 'A White Standard, embroidered with Silver and Gold: in the Middle a Bunch of nine Arrows tyed with a Wreath: all stained with Blood. The Launce broke, the Cornet killed without falling, being buckled behind to his Horse. Motto *'Alterius Jovis altera Tela'*.

The whole of the cavalry now dashed into the battle, clearing the lower of the wooded hills, driving the enemy into the Main on the other flank. It was now that the King, whose personal gallantry had been conspicuous, called off the fight. The Royals came off lucky, losing only three men killed. But there is a regimental tradition that their black plumes remember the old fight against the Black Musketeers of France.

* * *

The campaign ended tamely with the failure of the Austrians to force the Rhine while their allies looked on. A laborious march to winter quarters followed, the Royals marching with the Greys, Bland's Dragoons and the Black Watch. Next year was uneventful, though Britain and France were now officially at war. Generalship suffered a set-back when Wade replaced the sturdy old Stair at the head of the British, while the French gained immeasurably when the Marshal de Saxe took over their army (with the French Court holiday-making in his train). King George went home, leaving his second son the Duke of Cumberland in command.

Saxe had first to counter an Austrian invasion of Alsace and Wade could therefore move his immediate British troops down to live at enemy expense. Cumberland, the supreme commander, when he concentrated his forces, could only match 60,000 troops against the French 90,000. Under such circumstances the battle of Fontenoy, fought on 19 April, 1745, can only be excused as a piece of youthful eagerness. It was a battle as full of gallant episodes as of military mistakes, some of them near disastrous. The cavalry were left without orders for most of the day and when they did move it could only be to cover the retirement of the infantry. In this the Royal Dragoons gained distinction, crossing a hollow way under severe fire at a cost of 16 men killed, 36 wounded and a hundred horses killed.

Fontenoy may rank as a British defeat – though the regiments concerned applied for the grant of a battle honour before World War II – but it may not be unfair to repeat the eulogy of the victor de Saxe: 'Such things we have seen but pride forbids that we should speak of them, for we know well that it is not in our power to imitate them'.

The campaign petered out and the British contingent was reduced by the urgent need for troops at home, where Prince Charles Edward, son of the Stuart pretender, had set Scotland aflame and was threatening England. The British moved to their ports but transport was scarce and the weather discouraging, so that the Royals, under orders to embark at Wilhelmstadt in September only sighted the Thames on 1 December and took no part in quelling the last attempt of the Jacobites to restore the exiled dynasty. A week after they landed an eyewitness wrote of seeing them riding through the City 'and made a very fine appearance, the men appearing to be in good health and spirits'. Peace on the continent was not concluded until 1748 but the years were filled in by the usual round of home duties. In April, 1747, for instance, they had detachments at Hythe, Folkestone, Ashford, Lewes, Arlington, Houghton, Pevensey, Bourne, Arundel, Lydd and Romney, most of them assisting the preventive men against the smugglers – who were virtually the whole of the population.

* * *

The next war was never very far away. There was a strange uneasiness in European partnerships, the only constant factor being the enmity between France and Britain. The latter had now made friends with Frederick of Prussia who himself faced a threat from Russia, while the French had placated Maria Theresa, the disgruntled lady who had lost Silesia to Frederick and now spurned Britain who had sacrificed Cape Breton to get her better terms at the last peace.

Within the British army there was a notable stiffening of discipline and a flood of reforming regulations from the Duke of Cumberland who had lost his previous nickname of 'Sweet William' and was now 'The Butcher' of Culloden. The Royal Dragoons were made conscious of changing battle tactics by being given a light troop. This was the first part of the Regiment to see service in the new war, being included in a series of 'commando' descents on the French coast in 1758. It played an active part in the burning out of the harbour of St Dervan, near St Malo and in the later occupation of Cherbourg, although the British got bloody noses still later in the Bay of St Cas. The Seven Years' War which 'won Canada on the fields of Europe' now got into its swing.

Britain started on a modest scale but built its army on the mainland up to 20,000, not counting allies in British pay. They came under the capable command of Prince Frederick of Brunswick, that rare example of an allied commander of whom other allies could approve. The immediate aim was the protection of Hanover but this, incidentally, covered Frederick's realm. The Royal Dragoons missed the undeserved shame which British cavalry felt at Minden in 1760. The greatly increased cavalry component of which they formed part now came under the dashing Marquis of Granby to whose popularity so many soldier innkeepers were to pay tribute.

The Royal Dragoons came from quarters in southern England, a steady well-drilled regiment 550 strong, having left their light troop behind. Conservative British soldiers were apt to doubt such new-fangled notions. Early in April they passed through Bremen to rendezvous at Fritzlar, south of Cassel, where they were brigaded with the Inniskillings and Mordaunt's (later the 10th Hussars). The French advanced via Dortmund under de Broglie, with the allies getting the worst of the opening rounds. Ferdinand showed good steady generalship, though an early engagement at Corbach, covering

Cassel in the angle of the Weser and the Eder went against him. He retired almost at leisure – 'a very comical operation: being kicked, then turning about to snarl, show your teeth and then walk off again'. At any rate it worked, for the enemy had been taught a degree of caution where British troops were concerned.

Eventually Ferdinand coiled up for his come-back, his moves involving a night march over bad ground, evasion tactics, better timing than the communications of the period allowed and a dawn attack under his nephew the Hereditary Prince of Brunswick. But de Broglie made the mistake necessary to give his opponent victory – the detachment of 20,000 men to take up the line Ochsendorf – Warburg along the small river Diemel. The Royals were in from the start, covering the wheel of the dawn attack into battle positions, with the allied and some of France's best infantry racing for a crucial hill near Ochsendorf itself. A slight hesitation on the part of the French and the Royals struck. They scattered *Royal Piedmont* and then hit them again for good measure. 'Conway's Regiment of Dragoons' (ie the Royals), wrote an infantryman, 'charged them in flank so that they were entirely routed and dispersed'.

It was time for Granby, fuming back with the main body of almost exhausted infantrymen. Ferdinand let him loose and he covered five miles at a good round trot, his guns, much to the surprise of the experts, keeping pace. Granby's 'bald-headed' charge (he had lost hat and wig) completed the job and gave the Royals yet another chance. Well up in the pursuit they dealt firmly with the Swiss *Planta Regiment* who rashly crossed their path and then rounded up their private bag of 21 officers and 419 soldiers. Thus did they add 'WARBURG' to their battle honours.

* * *

The war as a whole languished but minor affairs were frequent. Clostercamp features on no British colour but it was a minor triumph for well-trained resolute troops. Ferdinand, finding de Broglie's attentions before Cassel embarrassing and having the broad plains of north Germany at his disposal made a sudden dart for Wesel, the Rhine fortress which the French had seized in 1757. The Royal

Dragoons came up with a reinforcement just as the French General de Castries took post about the convent of Clostercamp to ease matters at Wesel.

On 15 October, 1760 an allied force, the Royals leading, moved out from their positions near Theinburg to clear up the situation. The infantry battle went badly. The brigade of British (23rd, 33rd and a grenadier battalion) fired every round they had, including those taken from the pouches of their dead before they drew off sullenly. Then Elliott, the cavalry brigadier, got his turn. The Royals and Mordaunt's caught *Normandie* off balance, in the open and winded. The horsemen inflicted 700 casualties, taking a colour and two guns, giving the foot time to rally. But Ferdinand had only been buying time and went back now to winter quarters.

Sickness did more than the French, so that by November Granby's hospitals were crammed and he was 1,200 remounts short. Ferdinand tried the desperate expedient of a winter campaign but it failed in the desert which Hesse Cassel had become. When he was re-inforced in the spring he was still very much outnumbered. In this story of a regiment the name of Lt-Col Johnstone should be remembered. He had been given a staff job and Granby offered to relieve him of regimental duty!

'I took the liberty of declining', he wrote, 'preferring the honour of commanding the Royal Dragoons to any other while my health will permit, as long as His Majesty pleases to continue me to them', – which goes very far to explain the spirit of the British Army.

Ferdinand's patience was rewarded by the largely infantry battle of Vellinghausen but the wear and tear of a long war was taking effect so that by November the Royals were down to just over 260. But Ferdinand was successful in his main objects, the defence of Hanover and the holding of territory which could be used as bargaining counters by the politicians. Lt Goldsworthy of the Royals had the right of it when he wrote: 'Our getting this ground will not, I daresay, be much talked of in England as no blood was spilt, yet we look on it here as one of the greatest manoeuvres ever made'. The year of Wilhelmstahl, 1762, was Ferdinand's and he was the outstanding example of a continental general of whom even the British could approve. On the opposite side there was such disruption that de Broglie was unseated by court intrigue. The end of the war was in sight and in March, 1763 the Royals, thirty years of peace ahead, marched through Holland to embark.

Little as it seemed to be appreciated at home, the British army was causing the military world to treat it with respect. A French officer remarked, slightly later: 'The excellence of the British cavalry is sufficiently acknowledged in Europe and its advantages consist less in the goodness of the horse than in the choice of the horsemen . . . The sons of rich farmers and tradesmen are very desirous of entering into the service'.

Chapter 5

WHEN the French Revolution burst upon Europe, Great Britain was inclined to shrug her insular shoulders. Although it was part of her creed that nothing good could come out of France there were plenty who agreed with the revolution in principle. The British felt that they had no right to raise horrified hands at the murder of a king, though their own monarchy had become a venerated institution in the person of King George III. But the letting loose of a tattered rabble across the Low Countries was another matter, with the old threat to what were regarded as British ports of entry to a trading continent. What we denied to Louis XIV and Louis XV could not be surrendered to the assassins of Louis XVI. The matter was settled anyway by the French themselves: they declared war on Britain on 1 February, 1793.

* * *

Pitt had pushed the policy of appeasement to the limit: nation and army were equally unprepared. Sir Henry Bunbury, who left an account of the times from the military side wrote: 'Our army was lax in its discipline, entirely without system and very weak in numbers. Each colonel of a regiment managed it according to his own notions or neglected it altogether. There was no uniformity of drill or movement: professional pride was rare; professional knowledge still more so. Never was a kingdom less prepared for a stern and arduous conflict'. Speaking of the first landing of troops at Wilhelmstadt he said: 'Here were the colours of the British first waved in defiance of republican France and thus began the great war'. But the men behind the waving banners were few – 1,700 Foot Guards and a few gunners bundled aboard such colliers as could be found in the Thames. They were soon reinforced by a scratch brigade of cavalry and two of infantry but even so, when they moved to face the foe two regiments were left behind as being 'unfit to appear in the presence of an enemy'.

The next reinforcement was the Duke of York, second son of the King, appointed to command the mixed force of British, Hanoverians, Dutch and Hessians, under the overall command of the Prince of Coburg, whose Austrians and Prussians had a well-fought battle to their credit. It was an unfortunate set-up. The Duke had studied on the parade grounds of Frederick the Great, who had indeed died as the result of excitement engendered by his last review, given in the Duke's honour. Frederick's system should have died with him for revolutionary France ignored it. The Austrians lived still further back in history. An early advance into France could have been decisive but a siege – in this case the siege of Valenciennes, looked more dignified. Without it Paris could have fallen and Napoleon might have pottered out his life in some provincial garrison.

When war broke out the Royal Dragoons, almost inevitably, were on coast duty in the west with a strength of little over 200 men. They at all events escaped drafting and set about their authorised augmentation with time to spare. Ordered at last to provide a field regiment of two squadrons, they were able to do so and still leave a substantial depot behind. There had been many changes since their last campaign, including the setting of the final seal to their claim to be full members of the cavalry club. To the lay mind it might be trifling but to a proud regiment it meant much that their 'drummers' became 'trumpeters'. That strange British military habit of the docked 'Cadogan tail' had also been dropped and horses were able to use their tails for the purpose for which they had been provided – the whisking off of flies.

One curious figure lingered on the regimental stage, a relic of the Seven Years' War in which a certain Thomas Garth had served as a cornet, becoming lieutenant in the year following and subsequently colonel. His non-military interest lay in the fact that, though a 'hideous old devil', he was supposed to have had an intrigue with Princess Sophia, a daughter of the king. She had borne a son of which Garth was almost certainly the father but who was credited by the more prurient-minded to the Duke of Cumberland, her unpopular brother.

* * *

Britain had in some measure made good her ominous start to the war so that, by mid-1793 two cavalry brigades lay along the Thames ready to embark. The Royal Dragoons, together with the Blues and Inniskillings, sailed from Blackwall on 10 June, with the Greys and the King's Dragoons to follow. What was left of campaigning weather was spent in marching up and down the Flemish dunes, with a threat at Maubeuge. The French army was stiffening, finding that revolutionary fervour was not enough. The Royals, possibly from the very fact that they were used to fending for themselves in their inhospitable homeland, came out of it well. They took the measure of the enemy in a neat action at Pont-à-Tessin where, together with the Bays, they caught some luckless French infantry in the act of forming square, killing fifty and taking a hundred prisoners.

During a winter back in Flanders bureaucracy reared its head to the extent that each cavalry regiment was allowed £20 for postage and stationery. Perhaps the Duke of York was demonstrating thus early that his talent lay in administration. Among his more practical achievements were a complete issue of new tents for his men and an extra forge cart per regiment. The British contingent was slightly increased and regiments brought up to strength for the campaign of 1794, which was marked by three brilliant cavalry actions, of which the Royals reaped the battle honours of 'BEAUMONT' (also referred to as 'Cateau' or 'Bethencourt') and 'WILLEMS'. With the 3rd Dragoon Guards and the Blues they made up one of the four cavalry brigades. Theirs was commanded by Major General Mansell, an uninspired general who rapidly lost the confidence of his men.

Coburg and his Austrians still treated every defended place as a fortress but the main point was that they were fighting on French territory – battlefields that were to become more famous 120 years later – and keeping the would-be invaders clear of the Austrian Netherlands. In strength the allies were gravely inferior to the French under Pichegru but the immaturity of the Republicans made this good, though the chances lost were not to be retrieved until 1815.

An allied advance towards Landrecies brought honour to the 15th Light Dragoons and the Austrian Leopold Hussars. At Villars-en-Cauchie two regiments of 300 sabres under General Otto charged six French battalions and a strong force of cavalry, backed by guns, into dire confusion. Decisive victory was missed through the dilatoriness of Mansell, whose troopers, 'with failing hearts and humbled looks', came up merely in time to witness the light cavalry reap what laurels they might. Mansell clubbed his brigade and brought the 3rd Dragoon Guards under enfilade fire which cost them nearly forty killed. The situation was saved by the Royals, who covered the retirement of the other two regiments: but to exasperated men it could only be written off as a first fumbling round. The Duke was mild in his dispatch, admitting that 'by some mistake General Mansell's brigade did not arrive in time'.

* * *

Two days later Mansell atoned for his mistake at Cateau, when the Duke's force was threatened from the direction of Cambrai by two columns under Chappuis. The Duke's dispositions were excellent and resulted in a red-letter day for his cavalry. By adroit use of ground Otto, with nineteen squadrons including Mansell's, formed unseen by the enemy in a hollow near Bethencourt and then advanced, while the enemy was held by feints made by both infantry and gunners.

Coming over one last ridge the cavalry saw their prey, 20,000 French infantry, *facing the wrong way*. Mansell's brigade was directed at the second of the French columns. The Brigadier, humiliated by the Duke having called out to the brigade in general while passing: 'Gentlemen, you must repair the disgrace of the 24th', was determined not to come out alive. Dashing ahead of his men he fell at once and was found later by his son, lying in a ditch, naked and with his throat cut. The French guns were outflanked by Austrian cuirassiers: the Royals swept through, sabring the gunners and dashing in upon the astounded infantry so that they broke across the open cornfields. The slaughter only ended when horses were blown and men tired of their grim work. The honour of 'BEAUMONT' was amply justified, although that same night, in pelting rain, the allies had to march to retrieve an ugly situation in western Flanders. By early May matters improved and the Duke of York was at Tournai.

On the 10th the British cavalry completed the brilliant trilogy of victories which made the year memorable. Now, under the honour of 'WILLEMS' the cavalry were really able to 'put on a show' with forces coherent and ably handled. Regiments available were the Blues, 2nd, 3rd and 6th Dragoon Guards, 1st, 2nd and 6th Dragoons, with the 7th, 11th, 15th and 16th Light Dragoons, whilst Mansell had been replaced by Ralph Dundas.

It was Beaumont with a difference, since the French were determined not to be caught napping again and the ground consisted of deeply furrowed rye fields. The Duke held an entrenched position around Tournai against which the enemy moved in two strong columns at first light. One of these was held off by the Austrian *Kaunitz* Regiment but the larger pushed on, drove in the British outposts and opened a destructive cannonade. Suddenly York, spotting a gap in the enemy formation, ordered up sixteen British and two Austrian squadrons.

Nine times they charged over sodden ground in a slow but steady advance. There were many riderless horses and it was remarked that the old troopers maintained their places in the ranks until snapped up by horseless men. Dundas' brigade, reinforced by six other squadrons, came into action as the French retired on Willems, a village across the Lille-Tournai road, with the Carabiniers (part of the brigade's reinforcement) eager to work off an old grudge against their French opposite numbers, whom they recognised still wearing their old Bourbon uniforms. Dundas' men hovered round the retiring French until one square broke, followed rapidly by the rest. It was the moment for which cavalry waited but seldom found. As the squadrons pressed home the squares dissolved and the heavies, now further backed by light dragoons, 'had the execution of the fugitives', taking 409 prisoners, killing a thousand and bringing off 14 guns.

After so much glory it is sad to have to record that the war fizzled out. The Emperor of Austria, who had made the campaign as an interested spectator, lost interest and went home, while the allied army withdrew deeper and deeper into his dominions in the Low Countries, where the ground with its innumerable waterways was impossible for cavalry. The British component withdrew across the Rhine and found themselves in the Münster area after a most distressing retreat in foul weather. The Duke of York was recalled and the Royal Dragoons ended up in Hanover, where they were to spend most of 1795. The war was virtually ended when a peace was patched up at Basle in the spring of the year, mainly on the initiative of the King of Prussia. The Dutch were disposed to be friendly towards the revolution and the Austrians refused to mourn after their share of the Netherlands. George III was averse to leaving Hanover without British troops but pressure was brought to bear and early November saw the Royals at Stade for embarkation, at a strength of nearly 240 of all ranks, with 22 women, 11 children and 183 horses.

It is futile to try to hand out here blame for this grossly mismanaged war. The success of the new French armies cannot be ignored, with their amateur fervour so unlike the military minuets of the past. They had few rules, but a creed for which they were prepared to fight and whose benefits they were determined to extend to others. Yet the British had no reason to feel shame in their first battles against the revolution, long as they still had to wait before triumphing over it.

Chapter 6

WHAT was nothing more than a truce between Britain and France – and that merely in Europe – brought a long pause for most of the cavalry. Whitehall was still eager for battle, willing to consider the wildest of plans. Fortunately for the mounted arm, the scenes of such were generally so far away that the transport of all but the minimum of horses was impossible. Soldiers were more expendable and more easily stowed and the evidence of how they could be frittered away instead of being used to effect may be read in the returns. In 1794, for instance, there were 'killed or *died in the service*' 18,596 men, while in the two years following there were discharged 'on account of wounds *and infirmity*' 40,639 soldiers.

It was nearly fourteen years before the Royal Dragoons, now definitely classed as heavy cavalry, again faced an enemy. Heavy indeed they were, for when a cross-section of the Inniskillings (a comparable regiment) were weighed at the turn of the century, they rode at an average of 16 stone in review order and two stone more on campaign. There were reforms in arms, clothing and equipment: a uniform drill was insisted upon but there was little up-to-date field training.

'In England', wrote one officer, 'I never saw or heard of cavalry being taught to charge, disperse and re-form which, if I ever taught a regiment one thing, I think it should be that. To attempt giving officers or men any idea of outpost duty was considered absurd and when they came abroad they had it all to learn'. The entire lack of higher formations prevented any general officer from acquainting himself with any other duty than what he could learn from the book – and few of them were reading men. The one thing the British cavalry had before all others was speed, mainly due to their superior cattle. Once loosed they were uncontrollable by friend or foe, the men had some knowledge of elementary sabre work, but since it was usually learned on foot, when they rode into battle they were apt to slice off their horses' ears.

* * *

The Royal Dragoons, having been alternately reduced and augmented almost at the dictates of the daily news, did a spell in East Anglia and then marched for Edinburgh. A subaltern of the day found life pleasant enough, with a first parade at 9 am, quickly turned over to the sergeant-major, while the officers went to breakfast. Field days were a joy on the hard Leith sands, 'having a most beautiful sea view before you are adorned by the shores of Fife'. The young man was anxious to learn but was told by the sergeant-major that he need not be too exact for 'if I give the words of command the men would know what I mean'. It was at this period, incidentally, that regimental sergeant-majors arrived in the cavalry. They were not yet the towering figures of army tradition but a cross between drill sergeant and adjutant's help, junior to the troop quartermasters.

In 1807 whilst the Regiment was in Ireland, the war now took a different turn – against Napoleon rather than against the French. This was to the taste of the British people who have always preferred a personal foe, from Louis XIV to Hitler. But things looked grim. An army had been shattered at Corunna with its leader, General Sir John Moore, killed and Wellington with Britain's very last army was maintaining himself with difficulty on the Tagus. It was thus that the Royal Dragoons were shipped from Cork to Lisbon in mid-1809. Their brigadier, Major-General John Slade, had been their commanding officer but was hardly a dyed in the wool Royal, since he had transferred from the 10th Hussars as lieutenant colonel. But he was popular in the Regiment to the extent that the men now sailing under his command subscribed a hundred guineas for a sword 'in testimony of their esteem and regard'. The esteem was not shared by many in the cavalry – he had already tried his hand under Moore without conspicuous success. At Sahagun, where the 15th Hussars had scored decisively against the French cavalry, he failed lamentably to support them with the 10th as he had been ordered.

His attitude to war was more fitted to the theatre than the

battlefield. He made a stirring speech to his astonished men, ending with the exhortation: 'Blood and Slaughter – MARCH'. The advance was halted twice until the length of his stirrups was to his satisfaction so that, by his own account: 'We arrived at the hour fixed but the French were gone'. Wellington, who had reservations about his cavalry anyway, did not view his new arrival with any enthusiasm. But at any rate Slade looked after his own. On one happy occasion in Spain he stood guard in person over a large supply of wine, warding off all comers with a stern: 'You must not touch this. It is all for the Royal Dragoons'. The regiment must have had a jolly evening, since an independent witness estimated that it was 'enough for an army'.

But Wellington was well content with the Royal Dragoons, reporting home: 'I think that I have never seen a finer regiment. They are very strong, the horses in a very good condition and the regiment apparently in high order'. He himself was in a tricky position and many serving under him considered that he was merely waiting for an opportunity to embark the whole force. He was not well off for cavalry, having only three weak brigades but had not as yet operated in country where they could get a real opportunity. He lost two important fortresses this year – Cuidad Rodrigo and Almeida. But neither his own 'croakers' nor the French knew that he was solving Portugal's unemployment problem by setting 40,000 peasants to work at creating, out of an irregular ridge of rocks, the Lines of Torres Vedras.

He could afford to take no chances, opposed as he was by Marshal Massena, Napoleon's 'spoilt child of victory', a wily old fox whose handicaps were jealous subordinates and a campaigning mistress. Wellington stood to fight on 27 September, 1810, at Busaco, sometimes condemned as a political battle. But it heartened his own retreating troops, taught the French more respect and prepared the way for a come-back. Busaco was an ideal 'Wellington' position, with his main body out of sight behind a ridge whence they could pounce on their enemies just as they were blown from a climb up the hill. Here was certainly no place for cavalry but, when the retirement was ordered, Slade's and Anson's brigades formed the rearguard and skirmished the whole way back to the lines. They suffered mainly from weather and bad going, horses were worn out from stony tracks and bad standings, being shoeless and underfed.

* * *

Torres Vedras spelt destruction to Massena's army, though he gained some respite when, realising what he was up against, he withdrew to Santarem. The Royals had in their ranks one of those rare literary soldiers who let a few rays of light shine on long-past battlefields and bivouacs: James Smithies, who ran away from home in traditional fashion to enlist. He was at first in the Band but was now enjoying himself skirmishing in Portugal. Mixed into his professional notes and personal experiences he had the strange quirk of recording the exact number of religious houses in each station of his regiment. But in his first attempt to come to grips the enemy had the laugh of him.

During the careful advance in the wake of the retiring Massena, the officer on picquet ordered Smithies to investigate a French outpost. The trooper first challenged and then, receiving no reply 'thought I'd make the fellow either speak or run, so I charged him at full gallop and cut him in two, but great was our surprise to find, instead of a living sentry a dummy on horseback stuffed with straw'.

For the men in the lines, that winter was mainly a matter of combatting boredom with what entertainment they could contrive, from private theatricals to fox-hunting. Officers assumed a variety in dress never to be equalled until Montgomery took over in the Western Desert. One young gentleman dressed so magnificently that he was taken prisoner by the Portuguese, who were certain he must be a French general.

* * *

Massena eventually broke quarters on 5 March, 1811, and withdrew with Wellington at his heels through country that had been completely devastated. There were engagements at Pombal, Redhina, Foz d'Aronces and Sabugal, with a notable battle at Fuentes d'Onoro, where the Marshal made a belated attempt to go back to the offensive. The Royals had their own trials with Erskine – who was, to put it mildly, unbalanced – in command of the Light Division and Slade (temporarily) at the head of the cavalry. It almost seemed as if the pair of them competed as to who could blunder more. The Royals had a handicap nearer at hand in their commissary who, according to a regimental account, 'from the bottom of his heart wished well to the regiment' but failed miserably in the everyday matter of getting supplies up. 'Mules would pass by with rum and corn and biscuit for Bull's artillery and the 14th but nothing for the Royals'. But then commissaries, the civilian providers who feature so strongly in the military journals of the period, were almost an occupational hazard in the Penisula army.

Regular remounting was almost impossible and, of the black horses which were the pride of the regiment, 'one half of the beasts were sore-backed and the whole in perfect dog-condition. Their coats were long and brown and a parchment skin seemed every instant liable to break through and leave the ribs bare'. So bad was their condition that, taking advantage of a temporary lull, Wellington ordered the Royals back with other regiments to good grazing grounds.

Massena's abortive offensive opened on 2 May and did not at first involve the cavalry. Both they and the Light Division were now happier, their proper commanders being back where they belonged. On 5 May, with the battle in full swing, a French outflanking movement caused Wellington to pull out his Seventh Division, sending both the 'Light Bobs' and the cavalry to cover its withdrawal. Outnumbered though they were by two to one, cursing the curved sabres which proved so inferior to the straight blades of the enemy, Cotton's men had a day out. Poor though their horses were at this time, the French – never notable as horse masters – were even worse off. During this engagement it was a squadron each of the Royals and 14th who covered Norman Ramsay when he charged his two guns clean through the enemy ranks; Lt Trafford who took his squadron in to extricate a squadron of the 16th; a single detachment who charged a battery head-on and Lt Gunning (Gubbins to his friends) who captured the commanding officer of the 13th *Chasseurs à Cheval*. 'Gubbins behaved devilishly well', admitted the regimental journal. Four British cavalry regiments bear the blazon 'FUENTES D'ONORO'.

* * *

It was a long arduous campaign, with much skirmishing and a lost opportunity at El Bodon owing to a staff officer mistaking his orders. They did not get into winter quarters until the end of November. The Regiment was looking a little better now, with a consignment of new hats and overalls from England. Until then 'their turnout in marching order was the most genuine thing possible: to wit an old pair of yarn stockings with an old rusty spur at the heel, an old pair of shag breeches: the hat was superlative, not a vestige of its former shape to be seen'.

The new year was not a happy one for the Regiment. They missed the triumph of Salamanca, being detached at the time under Hill to the south: they had more than enough of Erskine and Slade wore out the liking they once had for him. He was a most nervous general for a cavalry brigade. No sooner was it announced that the French were in motion than an order was given to turn out, 'Jack running about everywhere crying: "Bridle up, bridle up. The first dozen men for God's sake. God damn you trumpeters: blow, damn you. Haste, haste, gallop. God damn you corporal, tell those fellows to turn out and never mind telling off".' He kept 100 men on picquet daily.

It was early in June of 1812 that the Royals had their most disastrous day. Hill made a feint towards Andalusia, Slade being ordered to advance against the enemy cavalry under l'Allemand. There was a preliminary outpost affair in which the British had the worst of it. Then was seen the result, not merely of mishandling on the spot, but of a long period of being 'messed about'. Slade put his whole brigade into a hand canter at the end of a pursuit which had already gone a dozen miles: held back from several chances and then, when the enemy was posted on favourable ground about Maguilla, sent two squadrons of the Royals, supported by the 3rd Dragoon Guards, into the charge. The horses were blown on both sides before it started but the Royals were the more eager and the

7   *An Officer of the Royal Dragoons*, 1832.

8 *The Royal Dragoons on parade at Manchester in 1843, with Lt-Col Thomas Martin in Command*

9 *A troop of the Royal Dragoons in 1848 turning out in review order*

10 *Inspection by the Duke of Cambridge at Brighton*, 1866.

11 *Officers of the Royals in Dublin*, 1858

12 *Officers wait their turn to face the perils of the riding school*

French turned tail. Both British regiments started in wild pursuit, when there came one of those extraordinary incidents of mob behaviour which defy explanation. The French general put in a small reserve to cover his retreat. A cry went up: 'Look to your right'. The troops, tired after the chase and in intense heat, feeling themselves mismanaged once again, wavered. Someone shouted: 'Threes about' and there followed the spectacle of two opposing bodies of cavalry running away from each other. The French recovered first but were too exhausted to make a serious attempt to molest the British on their eight miles back to Valencia.

Hill ordered an inquiry but Wellington knew where the real blame lay. 'The Royals and the 3rd D.G.', he wrote to Hill, 'were the best regiments in the cavalry and it annoys me particularly that the misfortune happened to them. I do not wonder at the French boasting of it'. The Royals lost heavily, especially in horses. It was September before they escaped the heat of Estramadura and marched the 160 miles to Madrid, followed by the difficult retreat when Wellington was forced to abandon the siege of Burgos. The result of a year's work is reflected in the figures. At the end of it the Royal Dragoons had only ten officers effective, with 307 other ranks and 172 fit horses. They spent most of their winter in Spain while Wellington reorganised his cavalry, for which he had now more use than formerly. He sent one heavy and two light regiments home and the Royals got 187 horses out of the deal.

To ease matters still further Slade himself was sent home and Erskine flung himself out of a window. Slade was replaced by Major-General Fane, of whose competence there was no doubt. 'The change', wrote one officer, 'I consider a good one, as do most others'. Perhaps it was symbolic for the regiment that it shed the last vestige of the eighteenth century, discarding the now shapeless cocked hats in favour of helmets with a long black mane.

* * *

Wellington now commenced the great advance which was to bring him to the French frontier and then to final victory on enemy soil. He opened with a wide turning movement via Salamanca, enlivened for the Royals by a smart isolated charge which brought them 140 prisoners. It was the more important as showing them back in form. Vittoria, when it came in June, was no cavalry battle, though Wellington has been blamed for not pushing the enemy retreat. The troops had their share of the immense booty and a delighted Smithies writes of 'casks of brandy, barrels and boxes of dollars and doubloons, wearing apparel, silks, laces, satins, jewelry, paintings, sculpture – some even had state robes and court dresses on'. There were wild scenes in camp but he was careful with his own loot, a collection of dollars, changing them into guineas which took up less room.

Pursuit, when it got under way, was difficult enough. The roads, normally terrible, were blocked by felled trees: rations, whose lack was so often a delay in any British force, were more scanty than ever. The French, despite their losses, were still hampered by stolen treasure and small actions became more frequent. The civilians came off worse. Smithies mentions one occasion when British, French and Portuguese were 'all plundering at the same time in the same house. They plundered in perfect harmony, no one disturbing the other on account of his nation or colour'.

At long last the British soldier was a hero even in his own country. It was now fashionable to look at least like a soldier and if possible to be mistaken for one. *The London Chronicle* observed that young bloods who had never worn a sword were buying ochre to imitate the suntan of veterans from Spain. The army was, as its own poet wrote much later: 'A thin red line of 'eroes when the guns began to roar'. There was more – much more – to the war after Vittoria but the cavalry had little to do but march while the infantry must 'stand to be shot at' on both sides of the Pyrenees. The Royals confessed to having 'the best of any epoch during the war', stationed from September, 1813 to February, 1814, at Villafranca, in good quarters and comparative plenty. Of wine and idleness there was too much, reflected in the court-martial book.

Crossing the Bidassoa on 11 March, the Royal Dragoons gave three cheers – and were in France, rejoicing at being free of Spain but having to exchange good wine for bad cider. At last Soult, who was in command of the French forces on the frontier, was forced to admit that his Emperor was beaten. In common with the rest of the cavalry the Royals marched for the Channel ports, reaching Calais on 17 July. Crossing to England, they made a pause at Newbury to pick up their depot. There was a joyous reunion at the 'Green Dragon' and 'never was a mess so numerous and well-served'. Then they marched for Bath, Trowbridge and Bristol. The usual reductions followed but the Royals, well under strength anyway, merely got rid of their undesirables. Apparently they made no mistakes. One discharged man joined a gang of thieves, got caught, tried and hanged, all within three months of leaving the Regiment.

Chapter 7

WATERLOO has been described as a trap for the historian – he has so much material, much of it contradictory. The army had now become so literary and so many rushed to record their own experiences, merely from the point of view of their own regiments outside the framework of the main story, that there was bound to be confusion. The Guards have Hougoumont, the King's German Legion La Haye Sainte: the Gordons tell how they charged clinging to the stirrups of the Greys, while the Greys in turn tell of how Sergeant Ewart captured the Eagle which was in fact one of two, both taken by the Union Brigade. For long it was the Greys alone who wore it as a cap badge until the Royals took it to replace their former badge, promoting it, as it were, from their collars. (Officers already wore it in their forage caps, by royal permission). It is related that, during World War I a distinguished Frenchman asked why an officer of the Royal Dragoons was wearing what appeared to be a French Eagle. Rather embarrassed, the young man told the tale.

'And you wear an Eagle still – one hundred years after you took it from us? I take that as the supreme compliment'. But as it happened the Royal Dragoons almost failed to get to Waterloo at all.

* * *

The return of the Emperor Napoleon from Elba was the bombshell of the century. News, with the improvement of means of communication – and particularly of the semaphore – was really becoming NEWS and the flashes to Paris started almost as soon as Napoleon stepped ashore. In France he seemed to be overdue, especially for the army the Bourbons had done their best to destroy. Old soldiers were driven to frenzy by popinjays in feathers and lace. In the ranks, men ordered to number off would cry: '*Seize; dix-sept; Gros Cochon*' in their estimate of the stout Louis XVIII who had come back from his exile in England. Now, successive numbers of the *Moniteur* changed their tone rapidly from: 'The Monster has escaped from his place of exile', down to its final triumphant: 'His Imperial Majesty the Emperor is at Fontainbleau and will this day enter his loyal city of Paris'. King Louis felt the fear of him from afar and did not stay to argue.

The one great advantage the allies had was that everyone of consequence was in Vienna to clear up the mess Napoleon had left behind him. Some of them had troops in plenty, if at a distance. England, true to form, had deplorably few but she had a commander-in-chief (also in Vienna) who was the one general of note Napoleon had never met and who had not the slightest fear of an encounter. He even refused to take the man he nicknamed 'Jonathan Wild' seriously, for he was the imperturbable British aristocrat of his age. Commander-in-chief he became by common consent – and was not envied in his job.

For Britain, not only was the normal reduction of forces after a war in full swing; she had been engaged since 1812 in a futile war with her former American colonies. This was now over and the troops (many of them old Peninsulars) had started for home. There was a strong British element (but no cavalry) in the garrison of the Netherlands which, if not always of first-rate quality, was the nucleus of the army destined to drive Napoleon to final exile.

London, when the bomb fell, was in a state of high excitement, protesting against the Corn Law Bill. Troops, particularly cavalry, were spread throughout the town and its suburbs. Lord Uxbridge, commanding the cavalry, was especially impressive, riding down to Westminster for the Lords' debate, in full uniform and surrounded by his staff. But, when the escape of Napoleon was made known it was 'as if the ground had opened'. Troops and agitators disappeared and in twelve hours the leading cavalry units were trotting down to the ports. The Royal Dragoons had lately been reprieved from a proposal to send them to Canada, to 'face the arrows of some damned copper-coloured savages', and were now fobbed off with the plea that their services against smugglers were more important than a final throw against the arch-enemy.

Then, on 21 April, 1815, came orders which transformed every face. The Regiment was ordered from Cornwall to Canterbury and augmented by two troops. Their leading squadron was at Ostend by 14 May, horses coughing ominously as a result of having to swim ashore. They mustered at Ghent at a strength of 460, having left four troops to form a depôt. By the end of the month the Duke of Wellington was able to parade for a bevy of distinguished guests 48 squadrons of magnificent cavalry, mounted as only the British could afford. The Royals were content with his passing word of recognition and a few words to Colonel Clifton. They were in the 2nd (Union) Brigade of cavalry, with the Greys and Inniskillings under Sir William Ponsonby – an alliance they still celebrate.

* * *

Wiseacres have blamed Wellington for many things, even in this his culminating triumph. He has been accused of holding his cavalry back instead of sending them to the frontier, but his main pre-occupation, with allies some of whom were dubious, was not with the holding of ground – which included an allied capital – but with the protection of his lines of communication back to England, a pardonable fault with British commanders. He was content to know, as he had known in Portugal, that he had the Royal Navy behind him. In the cavalry under the competent hand of Lord Uxbridge, he had his solid mass of manoeuvre. It had been planned to form the mounted brigades into divisions, but time forbade. The brigadiers were all old hands who knew his form, except Dornberg of the forward light brigade. It was his misunderstanding of the Wellington method that caused the Duke to be 'humbugged' by his adversary.

* * *

The general course of the three-day campaign can hardly be traced here but, when the alarm was given, the Royal Dragoons paraded at 4 am on 16 June, and headed south via Grammont and Ath. Liaison with the Prussians did not go too well. Their cavalry under Ziethen, who were responsible for a large sector of the front, let the side down badly, while Gneisenau the Prussian chief of staff had his hands full at Ligny, where Napoleon was winning his last victory. Wellington had a strong outpost at Quatre Bras which held off Ney until it could be reinforced. The Royals had fifty miles to cover over congested roads and spent the night in a barley field with horses saddled and linked in column, and until they sent out their own patrols the British did not even know that Blücher had been defeated and was in retreat.

It was obvious that Wellington must retire to conform, quite apart from the fact that his infantry had suffered severe losses. However he was not unduly worried, since he was going back to the position he had selected at the outset; if he needed a lead it was the line Marlborough had once marked out for a battle he was not allowed to fight. Back in the Duke of Richmond's study in Brussels, Wellington, taking time off from the famous ball, had drawn his thumbnail across the map south of Mont St Jean and remarked to his host: 'I shall meet him here'.

Two squadrons of the Royals were among the cavalry who covered the retirement from Quatre Bras, Major Radclyffe's dealing faithfully with a not-too-venturesome force of *chasseurs à cheval*. They had been through it all before, including the blood-curdling yells of '*Vive l'Empereur*'. In frightful weather they rode back to a night that was worse than a battle, going into bivouac near Mont St Jean, drenched and hungry but comforted by a chance-found dump of spirits. They did not leave their sodden camping ground until nearly 11 am next day. The two British heavy cavalry brigades of seven regiments formed in mass 400 yards behind the crest which was the key of the allied position. The Duke was fighting on a shoestring, for behind the heavies there were but a few regiments of allied cavalry. One such had stampeded to the rear the day previous, panicking the already nervy Belgian capital and rocketing the price of carriage horses to the crowds of British tourists who suddenly found that war was not all a blaze of scarlet and gold.

* * *

Various excuses have been found for Napoleon *not* having won the battle of Waterloo: he was a sick man; he missed his veterans and was without his marshals. The plain fact is that he was beaten by a superior general who fought an extremely personal battle with troops he knew – even in their failings – as no general before or since. Above all he pinned his faith to the Redcoat. 'Give me enough of him', he declared before it started, 'and I am sure'. Enough he had – but only just. 'It was', he said, 'a close-run thing'. On the purely mathematical side he started the battle with 67,000 men (25,000 of them British) against Napoleon's 74,000 and 156 guns against 246.

Reading his opponent's mind correctly, the Emperor sought out his right wing, but Hougoumont held fast throughout the day and Napoleon was forced to switch his main attack to the allied centre. Never having fought Wellington and ignoring the advice of those who had, he was deceived by the innocent-looking ridge behind which the allied troops mainly lay. He had, as he thought, disposed of the Prussians: one mighty blow up and over the slight slope would decide the war. His Guard had its dress uniforms in its packs ready for the state entry into Brussels.

D'Erlon charged with five infantry divisions massed after the French fashion. In the intervals and following were the cuirassiers, with light horsemen to the flanks. It was all according to the book which the old soldiers had studied long enough. The time was just after noon with the French swarming around La Haye Sainte. The British skirmishing line, according to practice, withdrew to the main position, which some of the weaker brethren took as the signal to go to the right about. Wellington had two decisive pieces on the board – Picton's Fifth Division and the brigades of heavy cavalry. The infantry, like their commander, were mainly old hands and the cavalry, having moved forward a trifle to escape the worst of the cannonade, sat impassive. In fact they were handled as one division, and Wellington had given their commander more scope than he usually did to subordinates.

Finally, with the line as brittle as a biscuit, Uxbridge held them back no longer. He galloped across the front of the Union Brigade

Every year from 1894, when he was appointed Colonel-in-Chief, until 1914, Kaiser Wilhelm II presented a Waterloo wreath to be hung on the Guidon. It was normally presented by the German Military attaché, seen in this picture taken on the Curragh.

Manoeuvres in the 'nineties.' A detachment of the Royals.

15. *With General Buller in Natal.* J. H. Stewart, 1900.

and warned Ponsonby to prepare to advance. Then he spurred on to lead the Household Brigade in person – a luxury no general can afford. Ponsonby jogged up to the crest, paused a moment until he saw the Household move, then gave a pre-arranged signal to Colonel Muter of the Inniskillings. The Union Brigade topped the ridge under Wellington's own eye and dashed into history.

They struck first at d'Erlon's infantry, catching them after they had crossed the slightly sunken road magnified by some writers into a positive ravine. None but the outside files of the French could use their muskets but the Royals lost twenty men in the forty yards or so they had to go before they could close. They rode through some of their own infantry to get at the French but so great was the shock that in a matter of minutes three divisions of the enemy were in full flight. The Heavies streamed after them with Picton's men rounding up those who hesitated. Picton himself, already wounded at Quatre Bras, fell with a ball in his head as the fight became a free-for-all.

The Royal Dragoons had their three squadrons in line, Captain Alexander Clark's in the centre. He had to pass through part of the 28th Gloucesters and then he saw, some forty yards to his left, an officer trying to escape with an Eagle, that of the 105th of the Line. Shouting 'Right shoulders up' to his men, he rode for the trophy, followed by his coverer, Cpl Styles. Lt Gunnings ('Gubbins' of regimental memory) though himself severely wounded, kept the squadron in hand, a task becoming increasingly difficult.

Clark ran the Eagle bearer through who, as he fell, let his colour drop across the head of the captain's charger before Styles could grab it. Clark tried to wrench the Eagle from its staff, meaning to cram it into his pocket but Styles, the careful NCO, cried out: 'Don't break it sir'. Clark handed it over and ordered his coverer to ride to the rear with it before he himself spurred on to rejoin his squadron.

Most of the two brigades were now completely out of hand: Ponsonby was killed and Uxbridge, caught up in the fight, was just one more swordsman. He admitted later: 'The *carrière* once begun, the leader is no better than any other man. I should have placed myself at the head of the second line'. One letter writer told of the enemy who 'fled as a flock of sheep across the valley quite at the mercy of the Dragoons'. The Royals, at first occupied with infantry, came to grips with the cuirassiers. James Smithies, whom we met in Spain, was still in the ranks on this supreme occasion. He describes 'some riders who caught hold of each other's bodies wrestling fashion and fighting for life. I did the same and got through as well as I could'. Wounded and taken prisoner, he escaped in the final confusion and was taken to hospital in Brussels.

The brigades, or parts of them, charged through to the French gun lines before starting on their way back, harassed by French lancers, succoured by Vandeleur who sent his 11th and 16th Light Dragoons to their assistance. They had driven back an army corps, coped with the Emperor's best cavalry and wrecked part of his artillery. Uxbridge told how the cuirassiers, though they advanced to the charge several times more, 'always did it *mollement*, as if they expected something behind the curtain'. There was little or nothing. The Union Brigade was now at half strength but were called upon several times more. When not actually engaged they, together with the Householders spread themselves out in line to make a better show, both as a deterrent to the enemy and an encouragement to friends. One allied regimental commander explained that his men had bought their own horses and would very much prefer not to loose them in what they considered a useless quarrel.

The Prussians had now started to become really effective, with old Blücher pleading: 'I have promised Wellington: you would not have me break my word?' Once again the Union Brigade charged against the cuirassiers. Both sides were by now so weary that they halted at thirty yards range and used their carbines. They were sick of it all – and Waterloo was won. Wellington and Blücher met at La Belle Alliance with the Prussian bands playing 'God Save the King' in honour of the grimy remnants that lined the road. Not for them to argue which ally had done what – Napoleon was off the map.

* * *

A few days later a coach rattled through London bearing Wellington's dispatch for the Prince Regent. An Eagle jutted out of each window. They had been dearly bought and the more precious for that. The casualties in the Royals were five officers and 71 men killed, apart from those who died of wounds later. Only 162 of 380 horses were still effective. The Duke, gracious if he could not be generous, 'took the opportunity of returning to the Army his thanks for their conduct in the glorious action of the 18th' before returning to 'real soldiering' by calling for a return on the subject of those who had 'quitted the ranks without leave'.

There was no more real fighting, the French, without imperial impetus, lacked their old attribute of a quick rally after a repulse. The Royals fetched up first at Nanterre, seven miles from Paris: but they were better off in a great barracks at Rouen, which they shared with the Scots Greys until they were ordered home. They mustered at Canterbury, where the officers refought the great battle over the table at Wright's. Ipswich, Scotland, Dorset and Kent – with one welcome trip to London – there was to be variety enough. There came a succession of colonels, all veterans of the great battle, with King George IV declaring benignly that it was 'essential for the service that the Royal Dragoons should ever be held by an officer of rank'. But the Regiment was far more concerned when the next monarch, King William IV, decreed that it must shave off its moustaches, which might in future be worn only by the Household Cavalry and the Hussars.

Chapter 8

For most present-day readers the Crimean War means little more than the Charge of the Light Brigade and the Lady with the Lamp. To understand Britain's entry into a war against Russia it is only necessary to read contemporary political history. Britain was already conscious of the threat of a Russian steamroller through Afghanistan to India and it now became evident that the Tsar yearned – as his successors today – to explore the Mediterranean, regarded as a private lake by both the Royal and the French navies. Russia's way was blocked by the tottering Turkish empire, with its capital at Constantinople. On what were blatantly trumped-up charges she advanced against the Turks on the Danube and won a resounding sea-victory off Sinope on the Black Sea.

These events and fears prompted an allied effort to help the Turks. The Emperor Napoleon III had a lasting urge to emulate his great uncle and was also attracted by the novel idea of exercising remote control over his army by the new electric telegraph. Despite his interference the French expedition was an efficient fighting force and much better administered than the British. There were also a little later, the Sardinians (referred to inevitably by the British soldier as 'the Sardines') an earnest of the new Italy yet to be born.

* * *

The Royal Navy also achieved a long-standing ambition to sail the Baltic, while the army appeared to be directed more logically to the task in hand. Although it had first to cope with the Russian threat on the Danube – a task for which the Turks themselves proved sufficient – its eventual task was the reduction of the Russian naval base at Sevastopol in the Crimea. The port was heavily defended and from the land side there were four rivers to cross from the point of disembarkation, including the Alma, which not only provided

Britain's first victory but sounded so attractive that parents used it for generations as a name for their daughters.

In the mid-nineteenth century it was rightly pointed out that Britain had no army, but a collection of regiments. To an extent this was its strength. Each regiment was fiercely convinced that it was better than its neighbours, although they were rarely on visiting terms by reason of their wide deployment. They were still the main support of the civil power, which was quite unable to support itself. Men knew their own officers, polished the numbers in their caps and fought for their own traditions rather than for any high-flown cause. There were neither formations nor services, all of which must be improvised for war. Spiritually the army was still under the command of the Duke of Wellington, who died in 1852. The very commander-in-chief designate, Lord Raglan, had been his military secretary at Waterloo and was apt to refer to his Russian enemies as 'the French'.

The Royal Dragoons, who had been almost continually on the move since their return from France, had just one Waterloo medal left and that on the breast of their colonel, Clifton who, now in his seventies, had little active influence. The Regiment was a little younger than formerly. In 1829 pensions had been introduced for men completing 21 years' service, so that the ranks were no longer clogged by veterans who just had to soldier on or starve. But the Waterloo men were still a vivid memory: 'Terrible old Ruffians; their language would have shaken the Devil off his throne and they drank whiskey like water'. There had been a sensation in the Regiment when new guidons were issued in 1852, the College of Heralds having disallowed the Regimental motto *Spectemur Agendo* which had been adopted from the Marlborough arms in 1740. But the Queen herself stepped in and the motto survived.

Reductions and poor recruiting had played their usual havoc and it took five regiments to make up the ten squadrons of Scarlett's Heavy Brigade which, together with the Light Brigade – also a skeleton – made up the cavalry division ordered east. To the original regiments of the Union Brigade were added the 4th and 5th Dragoon Guards. The training of heavy cavalry was still restricted to what was considered their rôle on the battlefield as the 'weapon of opportunity'. The heavy dragoon knew a little horsemastership, could drill to perfection and was familiar with the sabre exercise of five cuts and a thrust. For the past two years the men had been compelled to learn to jump their horses, a perilous sport for men who often rode at 21 stone. They were 'huge men on huge horses, with mighty accoutrements'. W. S. Gilbert was later to refer to 'the popular mystery known to the world as a heavy dragoon'.

The Royals embarked at Liverpool in May, 1854, at a strength of just over 300 commanded by Colonel Yorke. By the time they arrived at Constantinople it had been decided that the threat on the Danube was no longer serious and the army was directed to the Black Sea, the Heavies being delayed by lack of transport. The Royals were glad of the rest to recover from the journey out. Cholera and lesser ills were rife among the soldiery, though the Royal Dragoons were comparatively free of it. The horses had suffered most. Farcy and glanders hit them hard and thirty had to be shot. Towards the end of September the brigade embarked for a rough voyage in rickety ships towed by steamers. Such were the casualties among the horses that the Royals were only able to parade at all by taking over 75 from the Light Brigade. The deck of one transport collapsed resulting, as one officer wrote, in the 'wreck of as fine a troop of horses as ever I saw in the service'.

The allies were now committed to a full-dress siege of Sevastopol. The Russians had a formidable field force in being and the Balaclava position covering the main British supply base was made the responsibility of the cavalry. The 93rd Highlanders covered the village itself and Turkish troops manned a line of redoubts on the commanding ridge over which the Russians must come. Picquet and skirmishing work was arduous and the heavy cavalry had to learn new tactics. It was, said a regimental diarist, the 'poorest fun I know of', while another considered himself worse off than a London cabby since, 'if you keep a cabby out all night he charges you double fare, whereas a Heavy gets nothing extra but a cold'. But their day was near.

* * *

Day broke on 25 October, 1854 with abundant evidence that the Russians were preparing to attack the line of redoubts. In the first phase of the action some of these positions were overwhelmed and the Russians next advanced against the 93rd and their Turkish supports. The latter moved off, escaping the swords of the Russians but belaboured by the washing sticks of the women of the 93rd, who had seen no reason to vary their domestic programme. The 93rd stood firm and became, from that day on 'The Thin Red Line'. Lucan, commanding the cavalry, sent off the Heavy Brigade – less the Royals – to support them.

On the move to a new position Scarlett grasped his chance as a mass of Russian cavalry swarmed across his path over the ridge. He had a difficult manouvre to perform, a troublesome vineyard making it necessary to change formation twice but there was no doubt as to the drill of his squadrons. One of the Inniskillings and two of the Greys were first into line, with the 'Skins' second squadron following and the two Dragoon Guards regiments pounding up fast. Eight squadrons at an average strength of a hundred charged against sixteen squadrons of the 10th, 11th and 12th Russian Hussars, a regiment of Ural Cossacks and three sotnias of Cossacks of the Don.

The enemy helped in his own defeat. As an eyewitness told: 'The left wing of the Russian cavalry had shown first on the ridge, trotting fast. Their commander, General Ryjoff was on their left and they wheeled slightly until their centre was opposite Scarlett's and in depth equal to their frontage. They were about 2,000 strong with no artillery and no intervals. At between 400 and 500 yards they halted, then moved on at a hand's jog'. As Scarlett advanced the Russians halted again, fired from the saddle and came slowly on. Hesitation was their undoing but there was none on the part of the Heavies. Scarlett himself was the first man in, followed by his ADC, Lt Elliott, his trumpeter and orderly. The brigadier had five wounds, Elliott fourteen.

Yorke and his Royals fumed for a while then the colonel moved on his own initiative. The brigade was now closely engaged, the Greys cutting their way into the mass. Old ties strengthened the arms of the Royals as a voice cried out: 'By God, the Greys are cut off. Gallop,

gallop'! Then, in Kingslake's words: 'There broke from the Royals a cheer. Their trumpets sounded and, with ranks imperfectly formed, the Regiment advanced against the right flank and rear of the wheeling enemy mass'.

It was enough: the Russians started trickling back and then broke. The Heavies were this time well in hand and particularly the Royals. The total losses of the brigade in the charge itself – they lost more in the subsequent artillery action – were barely eighty. Nothing can detract from the glory won by the Light Brigade, enhanced as it was by Tennyson, but the very success of the Heavies helped to push their own achievement into second place, at all events in men's memories. 'In eight minutes', remarked the onlooker, 'it was all over and a few minutes afterwards the Heavies looked as if they had never been engaged'.

It had not been so quiet and orderly as all that. One survivor commented: 'It was just like a melée coming out of a theatre; jostling horse against horse, hacking and pushing, violent language, until suddenly the Russians gave way'. It is not unfair to add the comment of an infantry officer: 'The Light Brigade of over 600 sabres, after cutting their way thro' the Russian batteries and running the gauntlet back again, mustered only 300. The Heavy Brigade suffered little and did much execution'.

* * *

It was now that Russia's oldest ally – 'General Winter' – marched in against the wretched allied soldiers although even here the cavalry had the best of it. They at all events had exercise enough to keep them warm for, although they were disgusted at the indignity, their mounts were used as pack animals to bring up supplies over the quagmire that passed for a road up from the harbour at Balaclava. They lost horses in the process but some comfort was obtained by the building of 'do it yourself' huts to replace the tattered canvas tents. Relief was on its way in various forms, including gangs of navvies to lay a railway, and reinforcements arrived both in horses and men.

Funds raised at home sent out comforts which were sold at low prices and broke the infamous ring of extortionists who had flocked in from the Levant. True to their cloth, troops and officers still found plenty to complain about. The Regiment was particularly scornful about the quality of recruits and remounts but then 'things are never what they used to be'! There was also a notable imbalance between the number of horses sent and the men required to look after them – another normal complaint in the cavalry. But it was over at last, the technical victory of the reduction of Sevastopol being set against a background of untold and largely unnecessary suffering. The journey home was a distinct improvement on the outward trip of only a year earlier. The Royal Dragoons were now able to do the voyage complete in one steamer, the *Himalaya,* landing in May 1855 at Portsmouth.

Chapter 9

THAT the first station of the Royal Dragoons on return from the Crimea was Aldershot showed the shape of things to come. An almost unknown village was to become the home of the British Army – yet was not a direct outcome of the recent war. Thinking soldiers had already recognised that salvation lay largely in concentration at training stations and this was becoming practicable as the new civil police grew up and it was no longer always necessary to 'call out the military' and read the Riot Act in times of local stress. Many military stations still meant the splitting up of units but a start had been made before the war with the purchase of 10,000 acres at Aldershot for barracks and training grounds. There were developments also at such stations as Shorncliffe (famous as the training ground of Moore's Light Brigade), Colchester and at the Curragh near Dublin. The Army had a long way to go: it still had no regular administrative services – supply, transport and hospital arrangements were notoriously bad but had now been shown up to the nation. The personal attention of the Queen focussed opinion, so that the Crimea War was the beginning of the end for the out-dated army which had won Waterloo but must now step out, not merely trained but equipped as a modern instrument of war.

Regimental reforms were set afoot: an army reserve was initiated which at first served only the infantry but was extended to the whole army so that men, enlisting for twelve years, split that period between colour and reserve service according to the needs of their various arms. Cavalrymen signed on for seven years with the Colours and five with the Army Reserve and the whole system was graded down so that, when the Army Medical Corps was formed its soldiers stayed only one year with the Colours and the corps was thus capable of rapid extension for war.

The great reform of the century known as the Cardwell scheme of 1881 dealt largely with the infantry, the old mainly single-battalion regiments being grouped in pairs and given territorial districts, sometimes very much against their own wishes. The unwieldy militia provided reserve battalions for the county regiments, although it was left until the next century before the Volunteers, later the Territorials, could be absorbed into the system. It found its supreme justification in 1914, providing a readymade framework for Kitchener's armies. The almost feudal system of commission and promotion of officers by purchase was swept away. Its good points were almost accidental: it produced a very occasional Wellington and too many elderly gentlemen such as Burrard and Dalrymple to stand in his way. There were commanding officers who could hardly be prised loose from their emoluments and grey-headed juniors too poor to purchase a step and afraid of the penury of retirement.

One illustration from the records of the Royals may stand as an example – not a rare one. A most worthy NCO became cornet and riding master in 1832, lieutenant in 1835 and was still at the head of the subalterns' list in 1854 when, with the Regiment augmented for war he got his step without purchase, having seen twenty officers pass over his head. Augmentations and battle casualties were the only hope of the indigent, such vacancies being handed out free, not surprisingly the toast at the opening of a campaign was often: 'A long war and a bloody one'.

A perpetual plague to regimental staffs was the ever-changing establishment. There was, as might be expected, a rush to take from the Royals the two extra troops they had gained for the Crimea but they had to be raised again in 1856 for the Indian Mutiny – in which they were not employed anyway. The process was made more difficult as sixty men had already volunteered to regiments ordered abroad. In 1869 the old cavalry formation of the troop as an administrative unit was dropped, regiments now being divided in peace as they fought in war, into squadrons. There was an additional complication in that cavalry were assessed by weight, the Royals, Greys, Inniskillings and 4th and 5th Dragoon Guards being 'heavy' and therefore exempt from service in India. To such an extent was the practice carried that one CO of the Royals enlisted only six-footers and the regiment was nicknamed the 'Aldershot Guards'.

* * *

The Royal Dragoons did not even *look* like they used to. When, in spring of 1856 the Queen inspected them for the first time, they wore the new pattern helmet of white metal with its golden star and the black plume falling from a socket instead of being swept back like a mane. With minor changes this was to be the dragoon helmet to be worn (except in the Greys) as long as full dress existed. Cavalry drill was changed, too, the new 'section' being of four men, of whom three could dismount for action while the fourth led the horses away. Cavalrymen were on the way to becoming 'dragoons'.

Before the Royal Dragoons lay forty-three years of peace, divided mostly between Aldershot and Ireland, with a first five-year stretch in the sister island starting in mid-1856. Both stations now had a cavalry brigade in being, so that these two at least could provide an immediate battle formation trained and commanded as such. But, when the Royals came back to England in 1861 it seemed like the old order of things. There were periods – with consequent detachments – at Birmingham, Brighton and Manchester, none of which had accommodation for a complete regiment. Back in Ireland in 1867 things were even worse: civil strife was rampant and the Regiment had the unwanted pleasure of touring almost every county.

They were free of the long series of minor nineteenth century wars, though individual officers took advantage of any opportunity to get detached. The Regiment did however contribute to the ungainly Heavy Camel Regiment which was the main ingredient in the desert force designed to rescue Gordon from Khartoum. Regimental detachments were small – in the case of the Royals two officers and 43 picked men. They were not happy out of their regimental environment and hated the beasts they were given instead of horses. The detachment formed one corner of the square at Abu Klea, a gallant action for which no battle honour was awarded.

At home the new peacetime deployment was put on show in 1877, when a whole army corps and a cavalry division was reviewed by Queen Victoria in Windsor Great Park – the first time ever for such a massive parade in England. A year later there was a new threat of war with Russia – again to underpin the Turkish Empire. Most people deplored its existence but its disappearance would have robbed the western powers of some sort of a bastion. The army corps and the cavalry division were mobilised on a war footing and the army reserve was called out for the first time. As far as the Royals were concerned the new reserve was not a success, they received just 21 men of whom nine were medically unfit – the system not having got into any sort of swing. But there was no war and the Royals, returning to Ireland in 1880, found things sufficiently exciting, with 'boycott' now a dictionary word and snipers behind the trees. The posting meant six years of hard work without glory.

* * *

The appointment, in April, 1894, of Kaiser Wilhelm II as Colonel-in-Chief of the Royal Dragoons may have seemed decorative rather than functional but it was esteemed on both sides. It was not achieved without discussion, the Kaiser being the first colonel-in-chief in the history of the Regiment and the first foreign monarch to be gazetted to the British Army. Queen Victoria, on whom a parallel distinction had been conferred as Colonel-in-Chief of the *Garde Dragoner*, foresaw that the elevation of her grandson would be taken as a precedent both in Vienna and St Petersburg and thought that the Kaiser should be content with his rank as Honorary Admiral of the Royal Navy. Albert Edward, Prince of Wales, not always so affable to his nephew, thought otherwise and had his way.

With characteristic enthusiasm the Kaiser soon became a devoted 'Royal'. In the year of his appointment he slipped away from the Royal Regatta at Cowes and appeared in regimental uniform at Aldershot. In spacious days it was a small matter to bring a complete squadron of the Regiment from Ireland so that he could command it at a review. Every succeeding commanding officer paid a ceremonial visit to Berlin at the head of a deputation of officers. The Kaiser was an excellent regimental officer and encouraged tradition in his own army – not always strictly in accordance with history. Every year on Waterloo Day a German officer visited the Royal Dragoons bearing the Kaiser's wreath to be hung on the Guidon. The Royals acquired a sister regiment in the *Garde Dragoner* and formed an alliance with the *Kurassier Regiment Graf Gessler Nr. 8* of which the future King George V was colonel. The latter tie has been confirmed recently by the amalgamated Blues and Royals, with the 8th Cuirassiers at Cologne. Even after his abdication the Kaiser, whose appointment had terminated naturally in 1914, remembered his old regiment every year with a Christmas card.

On a lowlier but regimentally more intimate plane may be mentioned the retirement after 36 years' service of the Quartermaster, Capt H. L. Webb. His most famous *bon mot*, made to an inspecting general who asked him the secret of his success – which included not merely plain soldiering but fox-hunting and boxing ran: 'Well, general, I never said I couldn't nor I wouldn't nor as I didn't know how'.

But the Royals' long peace was nearing its end. It cannot be claimed that the army was fully modernised but in 1898 eight cavalry regiments trained together on Salisbury Plain. War was very near that autumn when French and British faced one another during the Fashoda incident in the Nile Valley. In 1899 there was a great royal review at Aldershot, followed by renewed training on the Plain. To the troops it was a matter for comparatively mild surprise that the manoeuvres finished ahead of time, the Royals being ordered back to Hounslow. But by mid-August they were holding themselves in readiness for South Africa, where they were to complete their training at the wrong end of Boer rifles.

Chapter 10

WHAT used to be known as the Second Boer War has become in modern terms the South African War of 1899–1902. Yet the term seemed fair enough at the time. 'Imperialism' had not yet become a dubious word and the British Empire was at its superb height. It was not even an exercise in colonialism, since neither side had the slightest regard for the original inhabitants. It was fought to decide merely which particular breed of colonists should rule the land. The Boers, in their efforts to escape the attention of British exploiters, had set up two republics – the Transvaal and the Orange Free State – loosely acknowledging the suzerainty of Queen Victoria. They had in mind the establishment of a United States of South Africa, for which the time was by no means ripe.

Britain made the appalling mistake of assuming she was up against an unruly collection of farmers. The Boers were neither 'unruly' nor a 'collection', but a nation trained and organised for war after their own fashion. Although Britain's forces on the spot had been raised to 25,000 in anticipation of trouble they were widely spread and the Boers were able to field twice that number with considerable, if anonymous assistance from Europe both in men and material. They were skilful as individuals and capable, when regular methods failed, of switching to real guerilla warfare. They scorned uniform and had the support of the country, together with interior lines of communication from first to last.

British troops were no less capable of dealing with them than any other European army might have been and had indeed shaken themselves free of the worst continental stiffness. The war might have been decided in a much shorter space of time had Britain been able to send out a mass of horsemen suitably equipped and trained. Before it was over it was necessary to employ yeomanry from home, horsemen from the colonies and regular infantry perched on ponies from Russia and Argentina. The winning weapons were not bayonet, sword or bullet but the blockhouse, barbed wire and the concentration camp, the invention of the last being used as propaganda by our enemies and later as an excuse for their own atrocities by the German Nazis. Yet the wretched Boer women who were rounded up and the cattle that were seized were among the most effective enemy weapons.

The army itself changed from a 'Thin red line of 'eros' to a band of 'gentlemen in khaki ordered south', for, together with its slowly modernised battle tactics it led the world in recognising that soldiers do not fight best in peacock attire. That much it had absorbed in Egypt and India. For the first time in a major campaign it also possessed a tail which really wagged – and that in a country where the pace of the supply column was that of the bullock wagon toiling over nearly non-existent tracks. On one occasion the Royals, marching over such a track, saw the box-seat of a GS wagon, with the whole wagon complete in the mud beneath it. They dug it out and used it as a mess cart throughout the campaign.

* * *

This time the Royals were left in no doubt that they would take a full share in the war and now for the first time they were able to call up their own reservists – 116 rather rusty but enthusiastic soldiers who, having rejoiced at being released from thraldom, were now delighted to find themselves back among their comrades. It was planned to reform the old Union Brigade and the Royal Dragoons embarked at Tilbury on 30 October, 1899 at a strength of 26 officers, 561 other ranks and 508 horses, under Lt-Col J. F. Burn-Murdoch, the only member of the Regiment with active service experience.

While the Regiment was at sea early experiences at the front flung a lot of plans into the melting pot, including those for the regenerated Union Brigade, which was never to parade as such. Four months before war broke out Lord Wolseley, commander-in-chief, had urged the concentration on Salisbury Plain of an army corps and a cavalry division but the home government was anxious to give no grounds for hostile comment so that, when hostilities opened and despite their 25,000 men in the theatre, the British force was still almost the

16 *Home from South Africa to fight against their Colonel-in-Chief, the Royal Dragoons now at Ludgershall, 1914, are seen wearing their 'Eagle' collar badge (instead of the authorized regimental cap badge) which eventually became the official cap badge of the Regiment.*

17 *Men of the 3rd and 4th Troops, B Squadron, at Vimy Ridge, 11 May, 1914, just after von Richthofen's Squadron had flown overhead.*

18   On service in France in 1918

19   'Jock', the regimental mascot, welcomed King George V and Queen Mary when they visited the Regiment in 1925.

traditional 'corporal's guard'. Boer enterprise had invested the garrison of Ladysmith and threatened Natal as a whole. The force of which the Royals formed part was therefore hurried round the Cape, the Royals landing at Durban and taking train for Pietermaritzburg.

Joubert, the Boer general to whom they were opposed, drew in his horns and took up a defensive position behind the Tugela River, giving the British time to organise, if on a provisional basis. Thus the Royals joined Lord Dundonald's mounted brigade, an enterprising Life Guards' officer who had come out at his own expense. General Sir Redvers Buller, a painstaking – and correspondingly slow – commander, organised an army of 10,000 men at Frere for the relief of Ladysmith and prepared to force the river line.

Now came the 'Black Week' of December, 1899, when Britain and her soldiers were made to realise at Colenso, Magersfontein and Stormberg the true quality of the men they were up against. Colenso was Buller's first attempt to cross the Tugela and the Royal Dragoons acted as a left flank guard to Hart's Irish Brigade. Almost everything went wrong, including the map which led Hart into a loop of the river which was not marked. The river was unfordable, Hart discounted three reports from the Royals as to enemy strength: two field batteries ran out of ammunition and there were severe casualties. Fortunately there were mistakes on the other side. Botha ordered a counter-attack which his subordinates were too inexperienced to undertake – in any case they were from the Orange Free State and resented the orders of a Transvaal general. The Royals suffered no casualties but the whole force had to be withdrawn while Buller turned his attention to the enemy right flank.

Bad weather now added its own handicap and it was mid-January before a force of three infantry brigades, six batteries and Dundonald's cavalry set off to secure Spion Kop, well up-river from Colenso. The stream was running high, pontoons were difficult to launch and the over-cautious programme gave the Boers time to concentrate. Hesitation, bad communications and enemy skill were preludes to a disaster of which the Royals were furious spectators. At one point the Regiment was ordered to occupy the decisive position but was saved by the sheer common sense of the commanding officer who, going forward on a personal reconnaisance up the hill, met stretcher bearers coming down with the news that a triumphant enemy was already in full possession.

Spion Kop was the Natal's army's last disaster but not its last check on the way to Ladysmith. Buller was perhaps typical of the old-style commander, cautious, much too methodical for the open spaces of the veldt; personally brave – he was the holder of a Victoria Cross – but now obviously old fashioned in his outlook. However he still held the confidence of his troops who, given their heads when the situation at Ladysmith became desperate, moved in almost stolidly to complete the relief by 3 March, although many declared it could have happened several days earlier.

For this concluding exploit the Royals were newly brigaded with the 13th and 14th Hussars and 'A' Battery, Royal Horse Artillery, with their own colonel, Burn-Murdoch, as brigadier. Covering the left flank of the advance, they swept round the Colenso position, with a sharp skirmish at Rustenberg. Although Boer opposition crumbled, there was still delay. The Tugela was treated as more of an obstacle than it deserved and the cavalrymen were forced to champ on eager bits when they might – at all events in regimental opinion – have been making for the real objective. The entry into the town was an anti-climax which gave neither the glamour nor the spoils of victory. Finally the Royals, with clean open veldt all around, were sent to camp on foul ground with consequent sickness both of men and horses.

For over twelve months the troops were employed on arduous duties in Natal, mainly on wearisome drives against an elusive enemy, with horse rations down to 4 lb of corn a day and no grazing. It was uninspiring work, though one such drive organised by Sir John French – now an outstanding figure among field commanders – rounded up 1,200 Boers and eleven guns.

* * *

The scene changed from Natal to the Transvaal, the Royals going by train to Pretoria on 5 April, 1901. President Kruger had decamped: outright annexation of the Boer republics had been proclaimed but the army had now to face a nation at bay. The Boers had no longer 'forces' but semi-independent bands fighting for survival. They had no supply services save those which were based on every farmhouse and staffed by their own women, who were in practice active enemies. The whole country was criss-crossed by wire, studded with blockhouses and swept by columns of exasperated troops. No one pretended that the concentration camps were places of comfort; they handed out primitive treatment to people used to primitive life – and they brought the war to an end, slowly.

For the Royals as for the rest, drive succeeded drive, starting in the Eastern Transvaal under Sir Bindon Blood. Home-bred horses were largely unequal to the task and the Royals brought only one survivor through, almost as a mascot. The rest, almost in the pony class, were bought wherever they could be found. For the troops it was, as one bored dragoon remarked: 'Full compliments and half rations'. But they learned from their opponents. They moved by night and grazed by day: the men carried tea, sugar and biscuits in their haversacks and took pot shots for their meat. The Royals worked under Plumer on the Olifants River; under Pulteney up to the border by Delagoa Bay and dropped from the snow line of the high veldt down to the orange groves. Back to Pretoria they rode to share in a chase after Piet Delarey and in an unsuccessful attempt to head off Smuts from Cape Colony.

* * *

The air now filled with rumours and Prince Francis of Teck had a useful friend in the person of Lord Kitchener. The two exchanged news over the telegraph in a rough code based on the Bible and Hymns Ancient and Modern. Prince Francis wired: '2 Kings 10.v.17', ('Is it peace?'), to which came the reply: 'Hymn 269, line 1', which runs: 'Christian seek not yet repose'. Resignation turned to triumph when Kitchener's log closed down with: 'Hymn 537', ('The strife is o'er, the battle done') which was followed almost at once by the departure of the Regimental detachment for the coronation of King Edward VII.

The Regiment itself had a three-months wait in Bloemfontein before embarkation for Southampton on 11 October, 1902, en route for Shorncliffe. They had had their fill of excitement, a glut of boredom, a minimum of glory. But they were now modern soldiers on a plane which no European army could match. 'RELIEF OF LADYSMITH' and 'SOUTH AFRICA, 1899–1902' stand for good solid professional work. If ever there had been glamour in war it went out when khaki came in.

* * *

King Edward was on the throne, the Empire at its peak. The army got back into scarlet and blue and green, the coronation ceremonies emphasizing the return to normal army life. For the Royals there was an impressive parade for their colonel-in-chief with a small shower of German decorations for services in a war which he had at times criticised sharply. Khaki did not as yet replace the time-honoured full dress, merely taking the place of red and blue serge which had been early concessions to active service conditions. Khaki had come from India and was therefore suspect in Whitehall. But training now ran along lines dictated by recent experience. A new cavalry manual shocked tradition to the extent of proclaiming that: 'The sword must become an adjunct of the rifle' – the final vindication of the 'dragoon' at the expense of the 'horse'.

South Africa showed also that the distinction between line cavalry regiments by weight was archaic. It was finally abolished in 1904 which meant in effect that the former 'heavy' regiments, including the Royal Dragoons, were placed on the peacetime foreign service roster. When orders were issued for the Royals to go east the only audible protest came from Germany, where the colonel-in-chief considered it undignified that his own regiment must now soldier in India. He was mollified by the reminder that King Edward's own particular pride, the 10th Hussars, had gone out to India as far back as 1845.

* * *

The Royal Dragoons entered into the new order of life readily enough. It is important to realise that the old army was essentially a collection of friends (and occasional enemies). Generations of officers came from the same schools mainly via Sandhurst: sergeants' messes were clubs which welcomed guests and canteens also played their part. Even generals were classified by the regiments to which they had belonged and were expected to live up to their reputation. Above all it was a happy, efficient if cursing army with a good-humoured contempt for the poor creatures who lived outside it. Everyone knew everyone else in a close communion of comradeship.

Five years at Lucknow were full of the normal incidents of military life in India and not without loss, as a memorial tablet with 49 names in the garrison church testified. This was the reverse of the medal, although India was free of even minor war during the stay of the Royals. Garrison routine was varied by honours to be paid to distinguished guests of whom the most exuberant was the Crown Prince of Germany, son of the colonel-in-chief. With two Kadir Cup winners placed at his disposal he wrought havoc – as one cynic claimed – among 'all the pigs in the United Provinces herded into a swamp'.

India could not always be made to fit into the framework of life accepted by a modern regiment. There was, for instance, that mysterious occurrence at Charaman, where the regiment was in camp. One afternoon, the horses of the entire Regiment, for no apparent reason, plucked up their pegs and stampeded over many miles of country. The Regiment, was, of course, furious but native wiseacres merely nodded and blamed it on to certain 'men of old' who were buried in the *bagh* behind the regimental lines. It was certainly curious that, years before, the 15th Hussars had suffered the same fate on the same ground.

The old regular army was headed for its zenith and it was appropriate that the Regiment should come under the influence of some of the leading soldiers of the near future, including Douglas Haig as inspector of cavalry. Horse, Foot and Guns, they marched on to disappear in their particular cloud of glory as the 'Old Contemptibles' – the product of the South African veldt and the stations of India. There was just one more bout of civil strife to cope with – and that in South Africa.

On 10 November, 1911 they embarked at Bombay. The Royals felt themselves at the top of their form – the new form of the new army. Not a subaltern in the Regiment had served in the Boer War and the rank and file were a new generation. But the torch had been handed down. Their first station was the comfortless one of Roberts Heights, where they celebrated – slightly delayed – the 250th anniversary of the raising of the Tangier Troop. A move to Potchefstroom was acknowledged to be for the better, until they were called upon for strenuous riot duty at Johannesburg in July of 1913.

The 'Jo'burg Riots' were among the ugliest instances of civil disturbance the British army ever had to deal with and the Royals shared the duty with the 10th Hussars. The trouble started in a strike among the miners and, as is often the case with strikes the men concerned were relatively peaceful. 'The people who looked like trouble', wrote one officer, 'were the real Jo'burg roughs, who are rougher and dirtier than any other roughs in the world', and behind it all was the real fear – not realised – of a native rising. There was the usual problem for the soldiers. Another officer wrote: 'They will throw bottles at us and if we retaliate we shall be prosecuted for murder and if we do not we shall be prosecuted for cowardice'. In the event they escaped both indictments though there were many among them who thought that patience was unduly stretched. In the end it was resolute action, including rifle fire, that won through.

Bottles were a fairly straightforward problem but in Johannesberg dynamite was in good supply and was made up by the rioters in handy form in bicycle pumps. The troops used their horses, the flats of their swords and eventually their firearms, after which the mob learned to discriminate between police and military. On one occasion a lone cab horse clip-clopping along a street was sufficient to send a crowd into a stampede but the troops, with cuts and bruises enough to show for their reward, were thankful enough when they were stood down and returned to their stations. The official report read: 'Their steadiness in the Market Square, the trying night firing in the streets, when troop leaders often had to act on their own initiative and the occasion in front of the Rand Club, were examples of the highest discipline'. In other words it showed two British regiments at their best. At the inevitable court of inquiry the damaged soldiers were produced with their bandages fresh upon them.

* * *

Time was running out. On the Waterloo Parade of 1914, when the Kaiser's wreath decorated the Guidon for the last time, the horse of the Guidon bearer bolted and had to be replaced – an omen men failed to appreciate. A little later the Royals were recalled from manoeuvres to be greeted by the news that the Empire was at war and that the Royal Dragoons no longer had a colonel-in-chief. They had been proud of the honour done to them but ceremonious pleasantries were past, grim realities close ahead. Not until 1922 was the vacant post filled, when King George V himself became colonel-in-chief.

The two great wars of the present century occupy so much space in our regimental histories that it is essential to remember that they were chapters and not books. So much more material was available, both private and through the media of public information. The Crimea War had let loose the dubious correspondent of *The Times*; the Boer War had released a flood of print, which became irresistible in 1914.

Chapter 11

THE Royal Dragoons had their own worry when war broke out. South Africa was too far away and there was the fear they might get entangled in minor affairs to mop up the German colonies with which the map of Africa was dotted, one of them on their very doorstep. Regular regiments felt that their place was by the side of their comrades in the British Expeditionary Force and not winkling enemy colonists out of dark continents. The Union authorities, however, undertook complete military responsibility so that all British units could be released for service in Europe and the Royals embarked at Cape Town on 23 August, 1914. Both they and the 10th Royal Hussars were detailed for the new 3rd Cavalry Division to be formed on Salisbury Plain.

With them went the native ponies to which they were accustomed. Some home authorities frowned but the Regiment was satisfied and many of the hardy animals saw the war through. As a first test of endurance they stood up well to a 28-day journey to Southampton and were warned for service in 19 days. The make-up of the 7th Infantry and 3rd Cavalry Divisions was almost scraping the barrel, with the government already nervous about home defence. The cavalry division could not even be completed. The Sixth Brigade had only the Royals and the 10th, with the 3rd Dragoon Guards on the way from Egypt. The 7th had the remnants of the Household Cavalry (who already had a squadron each in action) made up from surplus reservists of other regiments. The division was completed later by three Yeomanry regiments in advanced training stages. Ancilliary troops were in short supply, with only one RHA battery.

The sending of the two divisions under Rawlinson for the relief of Antwerp has come under a great deal of criticism. The addition of Mr Winston Churchill's private army, the Royal Naval Division (hardly out of its arms drill) was disastrous. But many soldiers had

20 One of the last of the 'Hanoverian Creams'. 'Coronet', the Drum Horse of the Royals (Sergeant Barnes up) in India, 1931. The horse was presented by King George V.

21 The Royal Dragoons are seen here doing traditional troop drill at Hadera, Palestine at the outbreak of the Second World War.

22 The Royals taking part in the victory parade at Aleppo after the Syrian campaign.

23 A Royal's patrol in North Africa relaxes beside the sea at Marsa Lucch.

24 'C' Squadron in Normandy passing through the shattered remains of Falaise.

25 In the snows of Holland on Christmas Day, 1944.

26 *The Royals pull up for the inevitable cup of tea, shortly after invading Germany.*

27 *The advanced guard of the Royal Dragoons pass through a small Danish town two days after the surrender of the German forces in North Germany on 6 May, 1945.*

always wished for a British force in Belgium and Rawlinson's men were at all events in time to form an effective left flank to the Expeditionary Force, now stretching out to defend the Channel ports. The 3rd Cavalry Division left Ludgershall for Southampton on 6 October. Mariners are always apt to be unsophisticated when required to transport soldiers, who are merely clumsy men with rifles with even clumsier horses. The odd business of cap badges has always been beyond them. Consequently, while one squadron of the Royals was shipped aboard the *Lord Charlemont*, the remainder were huddled into the *Indore* with parts of the 1st and 2nd Life Guards. It was inevitable that the two vessels docked as far apart as Ostend and Zeebrugge.

* * *

When they landed rumours were more plentiful than information. 'None of us know anything', wrote an officer, 'the Germans seem to be all over the place and our different armies all mixed up.' The enemy, as a matter of fact, were also reorganising and seemingly reluctant to engage until they could sort themselves out anew. The 3rd Cavalry Division moved forward to Bruges, deployed to cover the 7th Division, itself charged with succouring the Belgians in their retreat from Antwerp. The great harbour and fortress had fallen with comparatively little argument. Enemy field troops were not yet in strength and the obvious task was the link-up with the British 3rd Corps, now arriving from the Aisne together with Allenby's cavalry. The Royal Dragoons saw Ypres for the first time on 14 October, and had a sharp encounter the same day near Kemmel, to the advantage of 'C' Squadron.

Sir John French, possibly feeling himself at last free from allies he did not understand, seemed in an adventurous mood and planned a push back into Belgium. But the enemy was getting into his stride, with his main forces massing, the only people having any 'fun' being such irregulars as 'Captain Kettle' (Commander Samson) with improvised armoured cars. The Zonneback – Zandvoorde ridge became a defensive position which was to remain (with variations) throughout the war. The Royals dug their first trenches with tools from farms and lost their first officer, Capt Charrington. That losses were not severe was due to poor enemy marksmanship. The Regiment would have been sad had it realised that the day of the cavalry was over and that the main asset now was intensive training with the humble rifle.

The last days of October taught the cavalry what modern warfare was to be like. The Royals' position on the ridge had its strong point at Hollebeke Chateau and there were frantic days of give and take, with the line writhing like a snake. On the British side cavalry and infantry, including newly-arrived Indian troops, had often to be flung in piecemeal to restore the situation. On 30 October the Household Cavalry were decimated at Zandvoorde and the Royals held up an enemy advance until reinforcements could be trickled in. One day later they clung to their chateau as to a fortress, firing 350 rounds a man before, left unsupported because there were no supports anyway, they were ordered out. The division now formed the mobile reserve or, as one dragoon had it, became 'blooming night errants' – which a comrade improved into: 'Night and day errants'.

There was a typical example of the rôle that very day, when the Regiment stood to for yet another emergency. The roads were crowded with wounded, bringing well-founded accounts of an enemy break-through. Even the guns were pulling out when Sir Douglas Haig came up to order a counter-attack. Lord Cavan's guardsmen held an exposed flank from which the 7th Division had been forced back, Sussex and Northamptonshire men were at their last gasp when the last few of Bulfin's Gordons, with Cavan's own reserve of the 52nd Light Infantry were ordered up. The Royals arrived dismounted and got ready to take over the extreme left.

The traditional 'mad minute' of British musketry saved the day – and perhaps even the war, crashing into the startled enemy ranks as the Gordons charged home with the cheer that was the pre-arranged signal. 'Home' it was not exactly, since the enemy bolted, never believing that even the British would have taken such risks except in overwhelming strength. The line went forward half a mile before it could even be checked, with some of the Royals trundling 'C' Battery guns forward by hand. The Germans, in effect, never recovered from 31 October, 1914, though it was only the start of the battle of

Ypres and the weather now started to take a disastrous hand.

Cavalrymen were fast learning new skills: the defence of a trench line, the patrol into no-man's land, the collection of information by raiding parties on foot. Bayonets were issued – though Ordnance forgot they needed belts to hang them on. The Royals were forced back to their function as 'dragoons' in which the horses, often themselves searched out by fire directed from the air, were the means of getting to the vital point. 'It was not', wrote the regimental historian, 'on the specially critical days that the strain of the defence had been greatest', but 'to have to hang on day after day with every available rifle in the trenches knowing that there were no real reserves behind a thinly-held front line and being shelled from north, east and south simultaneously', – that was the test for which the Regiment had been moulded. This was not even plain patriotism, it was the tenacious regimental tradition that was the Royal Dragoons – that, and the sure knowledge that to the right and left were other regiments equally sure of their destiny.

Cavalry regiments remained 'regular' throughout the war and the same term, if technically incorrect, applied to the Yeomanry. Although so heavily engaged in the early days, the casualties of the mounted arm were not to be compared with those of the infantry, nor were they milked dry by the needs of expansion. When the cavalry needed drafts there were ample in their own depots, for the nation still lived effectively in the 'horse and buggy' age, and recruits came in readily.

* * *

The dismal story of the main stretch of the war need not be retold here, though the cavalry were rarely idle. There was this much to be said of the frequently criticised 'cavalry generals': they knew both when to use their men and when to conserve them. There was fierce fighting again in the May of 1915, when every man had to be sent up. Twice that year, forced by sheer necessity, first-class but not fully-trained men were used before they were ready. At Neuve Chapelle they were the best of the Territorials: at Loos the 'First Hundred Thousand'. In the final stages of the latter battle the Royal

Dragoons, together with the 3rd Dragoon Guards, were plugged in when the holding powers of the 15th Scottish Division were on the ebb. They stood to again for the tragedy of the Somme and moved up along the 8th Cavalry Brigade when it was mowed down by machine gun fire in front of Arras in the spring of 1917.

It was as an infantryman that 2/Lt John Spencer Dunville won a Victoria Cross for the Regiment on 25 June of that same year. The cavalry was holding a 'quiet' sector at Epehy. The Royals had just been relieved, leaving in the line a raiding party of a hundred men who were needed to investigate the German positions. Lt Dunville's bangalore torpedo, intended to do a vital job in cutting the wire through which the raiders were to penetrate, failed to explode and, while Sappers set about doing the job by hand the young officer, placing himself between them and the now angry enemy, opened up a minor offensive of his own. He was severely wounded but never lost control of the situation.

Another detachment under Lt Rice pushed through the outpost line, L/Cpl Jull clearing up three shellholes full of Germans with the bayonet. The raid was called off before final penetration could be made but in the meantime Lt Helme was killed, Dunville dying next day in hospital. The regiment next moved north to back up the ghastly Passchendaele effort, made necessary by the dangerous discords shaking the French armies. It is not unfair to claim that Britain's allies in that winter of 1917–18 were more of a hindrance than a help. The Russians defected altogether: the Italian front had to be bolstered up by British troops after Caporetto – at the expense of Haig's armies in France.

Early in 1918 the military authorities, despairing of a breakthrough in France, reduced the cavalry from five divisions to three, sending away the Indian regiments, dismounting the yeomanry and converting the Household Cavalry into machine-gunners. The 10th Royal Hussars rejoined the Royals in the Sixth Brigade and the 3rd Cavalry Division was now completed by the Canadian Cavalry Brigade.

* * *

The German offensive of March, 1918, had been anticipated but not its extent: it was the final effort and almost swept away the British Fifth and Third Armies. Reserves were scanty and rear positions sketchy. The 3rd Cavalry did a night march southwards on the first day of the attack and just after midnight a dismounted battalion from the Sixth Brigade, including a company from the Royal Dragoons, was sent to the flank of the retiring British line. Next day reports were alarming. The Royals' dismounted company, attached to a badly shaken 58th Division, hung on until withdrawn to Chauny and then again to the Oise Canal, the French on the right having gone out under cover of fog.

There was a red-letter day for a few of the cavalry on 24 March. General Jacques Harman, recently appointed to command the 3rd Cavalry Division, had assembled a force of 1,800 men, including a mounted squadron from the Sixth Brigade, each of its three regiments contributing one troop. With these he marched boldly into battle, though the historian admits that his infantry were 'a scratch lot and rather shaky'. With the enemy penetrating through several gaps, Harman's cavalry squadron now trotted up a sunken road near the village of Collezy forward of Chauny, halting behind a farm while officers spied out the ground. There were machine guns 600 yards to their front, while to the right were British infantry at the end of their tether.

The squadron leader, Major Williams of the 10th, sent the troop of 3rd Dragoon Guards charging to the right to relieve the infantry. Then the 10th, followed closely by the Royals, charged straight ahead. The distances were too long for the drill book but the need was desperate. There were 600 yards to go, the last 200 over plough. 'Knee to knee at first, opening out a little as they dashed forward, the Royals and 10th came right in among' the astounded Germans, from whose tidy minds horsemen and swords had long been erased. 'Weary British infantry who, five minutes earlier, would never have dreamed of advancing, forgot their fatigue and rushed forward to follow up the success'. The enemy loss was 70 killed and a hundred prisoners. The British gain – apart from the badly-needed boost to morale – was the rescue of the nearly-surrounded 36th Division. The

Royals lost six men, with three officers and six men wounded, Lt Cubitt dying of his wounds.

Still the enemy pounded on until he threatened the vital rail junction at Amiens. A major attack developed near Villars-Bretonneaux on 4 April, with a German thrust near Hamel. The Sixth Brigade filled the gap as, at this juncture, only cavalry could do. They came up at the gallop and were in action, dismounted, before the enemy could recover from this new surprise. Hardly rested from a gruelling period, the whole of the Cavalry Corps marched north, where the Portuguese were unhappy on the Lys. The situation restored, they went back for a relatively quiet summer.

* * *

August saw the real turn of the tide, with the cavalry grouped behind Fourth Army east of Amiens, the Canadian cavalry leading the division to co-operate with the 3rd Canadian Division. Again it was true dragoon work – advance at the gallop and dismounted action at the end of it. A series of blows forced the enemy back over his vaunted Hindenburg Line, with the cavalry active in patrol work. One real hardship as the year wore on and nights got colder was the increase in German bombing of bivouac areas, which curtailed sleep and prevented the issue of hot meals.

For the Royal Dragoons the war went out with a final dash. On 9 October, the Fourth Army put in its attack in the Le Cateau area with the 3rd Cavalry supporting 13 Corps. There was a check north of the main road to Le Cateau but General Harman ordered a mounted attack by the Sixth Cavalry Brigade in a dash forward from Maretz, led by the Royals. They went through the firing line and out into the open, followed by cheering infantry. Held up by machine-guns, the Royals swerved to a flank together with the 3rd Dragoon Guards, to threaten the now retreating foe. The divisional bag for the day included 450 prisoners, ten guns and over fifty machine-guns, all of which counted for less than the positive feel of victory in the air. The Royals had eight men killed, twenty wounded and something under forty horses hit. It was the last price they had to pay. Turkey had dropped out, then Bulgaria, then Austria-Hungary. The Royal

Dragoons moved across the field of Fontenoy, near the scene of their own triumph at Willems. The cavalry horse made his exit, earning his oats up to the end.

Armistice saw the Royal Dragoons still on the march, headed for Enghien. They were halted at Leuze with the news they were almost too benumbed to appreciate, returning to Ramecroix 'tired and miserably wet, with a strange feeling, part relief, part wonder that this was at last the end of the struggle that had seemed interminable'.

* * *

The Regiment reacted at once and automatically, spending a week getting itself clean, then over the field of Waterloo to winter near Liege, reaching Cologne on 24 March, 1919. It was now largely occupied with the trials of demobilisation, reaching Germany with a strength of less than a squadron but gaining 250 entirely unwanted horses from the 18th Hussars. Officers and men started coming in from detachments, drafts were received and the stay abroad was not prolonged. In September, 1919, the Regiment landed at Dover and were ordered to Hounslow, with its horses sent to quarantine at Luton.

Chapter 12

WHATEVER rôle there might be for cavalry in the future was for the government of the day to decide. The duty of the Royal Dragoons was – to become the Royals again. Re-forming the regiment was complicated enough. It was new from the top down. On April, 1919, General Burn-Murdoch became colonel, while the command went a few months later to Lt-Col H. A. Tomkinson, known far outside his regiment as an international polo player. Then, with the regiment recruited but far from trained, with frequent demands for men from regiments ordered abroad, came a posting to Ireland, troubled as never before. Many in the ranks were using their rifles in earnest before they had even fired on the ranges. They were engaged in guerilla warfare against an enemy prepared to take a drink with a soldier in a pub prior to shooting him in the back outside. The Royals were stationed at Ballinsloe but spread in small detachments as in olden times. They greeted the Irish treaty with enthusiasm, coming home first to Hounslow and thence to Aldershot.

It was in 1924 that the regiment enlisted yet another of its recording members, in Boy (later Trooper) 'Spike' Mays, whose 'Fall out the Officers' gives a graphic picture of life in a smart cavalry regiment between wars. Not free from the soldierly disease of grousing, he reveals himself as the good example of a regimental fanatic, glorying even in 'spit and polish'. His proudest claim is that during his service he was selected over forty times as 'stick orderly' – the smartest man of the barrack guard. His regimental education started on the way up to barracks from the railway station, when the conducting NCO told him the story of Waterloo and the Eagle. In a few weeks he was pitying his own brother, a mere Grenadier. He learned that he had no feet, but 'near and off hinds'. When he first went to stables he met thirty friends – all horses who never let him down!

King George V made his first visit to his Regiment in June, 1925, when it was at Beaumont Barracks, Aldershot. It was then he presented it with the new Guidon bearing the ten battle honours of World War I. The Colonel-in-Chief saw 'not a man and not a horse out of line'. But later 'the Bird-catchers' (a reference to the Eagle) had their own celebration of a great occasion.

'It was a night of toasting', says the recorder. 'We toasted our new Guidon, next our horses and then almost everything in cavalry sight ... Old comrades toasted the rookies and *vice versa*. Officers toasted other ranks and also *vice versa*.' Their Majesties, in their visit to the barracks were also greeted by Jock the regimental mascot, a goose who, having escaped the cook's knife in Ireland, had taken over control of the barracks square and eventually died in Egypt. But the Royals were given little rest. Late in that year they were back at Hounslow and sailed for Egypt in September, 1927.

There is no such thing as a 'typical' British regiment, since the whole structure relies on its very diversity but to visit the Royals in barracks between the wars was to sense the spirit that had endured for centuries, a loyalty to their own ideals. They took it for granted and had no need to parade what was obvious. All ranks knew where their duty lay and if they were at times puzzled from above they gave no sign of it. If they knew that the horse was headed for oblivion they lived and worked and trained as if they would ride on into history. They read the blazons on their Guidon with the quiet unassuming confidence that, if further honours were to be won, they would be gained without fuss. It would be what they had joined for.

Egypt was normally a halt on the way to India and was also 'in the day's work'. It was there, while dealing with the intransigence of the Khedive, that the Royal Dragoons became gravely concerned about the health of their colonel-in-chief. The BBC was regularly jammed and telegraphic communication was uncertain. The Royals got their news by visual signals from long distance until the message came: 'His Majesty's health is improved and he is resting quietly'. Then two lowly signallers stood to attention on the 'top rock of

Cheops and drank His Majesty's health in chlorinated water'. Of such, too, is regimental history made.

It was in Egypt that the regiment formed, with the Monarch's approval, their alliance with the Royal Canadian Dragoons, who had fought at their side during the difficult days of 1918. This regiment had been formed in 1883 and saw distinguished service in South Africa, where it was awarded three Victoria Crosses and a DCM for a running fight in covering an infantry withdrawal between Lilfontein and Belfast. One of its badges is the Springbok remembering the day when a particularly active animal had warned them of a Boer ambush. Their most recent exploit had been the charge of 30 March, 1918, which stopped the German advance on Amiens. They later went mechanised to Italy in World War II and took part in the final stages of that war in Holland.

* * *

Only a few very old soldiers had previous service in India when the Royals went there in October, 1929, and had hair-raising tales to tell of the North West Frontier – where they had never been! Times were difficult, with civil disobedience rife and soldiers taken far too often as visible signs of oppression. But there was to the end no bad feeling between British and Indian soldiers. Their own professional creed cut clean across all divisions of race, religion and politics. Symbolic of it all was the Gurkha soldier who, after a 'soldiers' night walked back to barracks with friends of the Royals.

'Halt, who goes there?' came the challenge of the sentry. The little man in bottle green stood to attention, head high as he answered: 'We are soldiers of King George'. The answer was as prompt: 'Pass friends . . . All's well'.

India was still the supreme training ground of the army. There was a minimum of restrictions: regiments were kept at war strength so there was no wastage of manpower for work that could be done by cheap labour. All ranks had of necessity to be security-minded: sport of all kinds flourished. Strange rumours came from England but the cavalry in India remained unmoved. The tour ended in 1935 but there was a short call in Egypt on the way home, where it was thought a stronger garrison might discourage Mussolini. By May of 1936, the Royals were at Shorncliffe. Mounting world tension was playing havoc with orderly peace-time routine and the Royals got an unexpected move. Instead of going quietly to Colchester as ordered in the autumn of 1938, they were shipped out to Palestine for the tricky job of umpiring the conflict between Jewish settlers and the original Arab inhabitants.

Chapter 13

Few regiments can have bettered the record of the Royal Dragoons for the second world war, either in length of service abroad or in geographical scope. They were 'at war' when it started and ranged over Palestine, Syria, North Africa, Italy and North West Europe to Denmark. They were abroad, with a short break, from 1938 to 1950. During the whole period they were strictly a regular regiment, taking in their first national servicemen in 1947. When the war started there remained in the army only the Household Cavalry, the Royals, the Greys and a few Yeomanry regiments still horsed and it was decided that, *if* there were scope anywhere for men on horses it would be in the Middle East. Consequently a whole cavalry division was built up in Palestine, though all ranks felt in their bones that the horses must go.

The British soldier is from long experience a philosopher and the outbreak of war made little difference to Palestine duties – mainly the searching of Arab villages – with pleasanter interludes on stony football pitches and warm bathing. The honour of still being horsed was a positive impediment to progress. Vehicles, especially armoured ones, were in too short supply for any to be spared for units tucked away out of sight, although the Royals had the advantage of having brought out with them from home a complete mechanised squadron of the 5th Inniskilling Dragoon Guards. The combination of vehicles and horses proved most effective for the duties in which they were employed and gave the horsemen some idea of how they would have to work in the future.

Just prior to the war they crossed swords with a future firm friend, a general who hardly appreciated the still considerable talents of the horse. 'Montgomery is the gentleman's name', wrote the commanding officer, 'and I am told he is one of the brains of the army'. To gain their point the Regiment took the general's GSO 1 on a raid in the Plain of Esdraelon. A cold night raid of 27 miles over rough country, followed by a successful pounce on an Arab village convinced the staff officer, impressed the general and caused the Regiment a few chuckles.

It might have been *'magnifique'* but was unfortunately not *'la guerre'* and in 1940 the Regiment made its own plea to be mechanised, selecting the armoured car as being appropriate for twentieth century dragoons. News from Egypt made them eager but there could be no short cut and the Regiment went back to school near Cairo – which was a home more often dreamed of than visited during the coming campaign. Their first sight of a real enemy was a crowd of docile, rumpled little men – Italian prisoners whom they escorted. Almost inevitably one of them came forward as an interpreter an ex-waiter from a Soho restaurant.

The Royals were lucky in the instructors posted to them: they came mainly from the 11th Hussars, old hands in this strange new trade with desert action behind them. Faced by necessity, spurred by ambition, the Regiment moved fast. They had their last horse inspection just before Christmas in 1940: their leading squadron was heading for the Western Desert four months later. They were, as a senior officer wrote later: 'trade tested by the Germans'.

Their instructors were probably responsible for the Royals adopting their new headgear, the grey beret, for the 11th prided themselves on being 'different' and were renowned for their brown beret with its maroon band. As long as there has been a British army its regiments have contrived to please themselves in details of uniform whatever authority have had to say. Colonel Heyworth now decided that his Regiment would look smart in grey berets, so grey berets – made in Cairo at regimental expense – they wore from now on.

The Regiment was trained and ready for war – more ready than many who were sent into the desert. They got used to the main exit from civilisation, the road 'up the blue'. The tightly-knit regiment of horses and men in built-up Europe became a ten mile column of vehicles 'festooned with such an assortment of kit, petrol cans, packs, sand-trays, strips of furled canvas that they had, at first sight, an air of irresponsible gaiety with billowing white dust such as a flotilla of destroyers might have thrown up in a heavy sea.' Equipment was still scanty: only troop leaders had wireless sets and the Marmon Harrington amoured cars had little but their stout engines to recommend them. At first it was largely patrol work they were given, troops meeting their neighbours two nights in five, with petrol more plentiful than water and enemy air more venturesome than his land forces. But there was another war which now intervened.

* * *

French-held Syria had not been expected to present much of a problem. But the troops in possession were regulars, sullen at the rapid fall of their homeland and ready for a fight whoever it might be against. Loyal after the manner of their kind, they regarded the Free French who took part in the march against them as traitors. One squadron of the Royal Dragoons had been left behind on the Syrian frontier and was later reinforced by a second, together with regimental headquarters. The attacking force in this 1941 sideshow was at first distinctly *ad hoc,* including Free French indulging in a private war, Australians and British of all arms, from regular infantry to Lifeguardsmen escaped from Whitehall in a strange assortment of vehicles. It was a troop of the Royals who escorted a Free French cavalry commander on his formal entry into Damascus; a squadron which covered the advance on Homs and a troop which blew up a vital railway bridge one night, only to find it was urgently required – in good repair – next day.

This particular war ended with the Royal Dragoons patrolling 'some of the most beautiful country in Asia Minor' bartering tea, tobacco and empty petrol cans for eggs and chickens. The British soldier has always been an expert linguist, usually without knowing one word of a foreign language. There was something for everyone, up to the officer who found the countryside 'reeking of men-at-arms and chainmail' from Crusading times. The squadron left in Africa, with worn-out vehicles apt to break down at the most inconvenient moments, was thankful to rejoin the Regiment in Syria on the last day of October.

Naturally it was too good to last: Syria was mollified, and the

Turks affable if persistently neutral. The battle was on in North Africa and thither the Royal Dragoons went at the end of November, 1941. Desert war brought the day of the private dragoon and the young troop leader and, among other things, traps for the map reader. Some trained on the theodolite or the naval sextant but most found their hand compasses, backed by 'the highest degree of skill and level-headedness' sufficient to bring them to Tunis, via Tripoli.

Battles were often personal, as when Cpl Owen knocked out an enemy armoured car, chased its companions until brought up short by a row of guns which put paid to his vehicle and wounded his driver. Owen dragged out the wounded man, went back for his first aid box and then lay up in the scrub all night listening to strange tongues. In the morning he reported for duty to the first officer he met – his own divisional general 'Strafer' Gott.

Forming part of Brigadier Jock Campbell's support group, the Royal Dragoons made their first real scoop. Campbell was ordered to 'put a stopper in the bottle' of Benghazi and headed across the desert. Lack of petrol grounded much of his group but 'A' Squadron of the Royals fell in with an Italian petrol lorry. Thus it was that Major Pepys, with his squadron, plus a company of the Rifle Brigade, a troop of the Northumberland Hussars and a few Sappers thrust on. Eventually 'Pepforce' entered Benghazi on Christmas Eve, accepted the formal surrender of the city from the Mayor at 11 am and were able to report for the war diary that 'the KDGs arrived about 1600 hours, preceded by the press'. It was far from being the beginning of the end and, from the regimental point of view there was a grievous loss before the month was out. Colonel Heyworth, who had brought the Royals through the perils of mechanisation as a most popular commanding officer was, with his second-in-command, the victim of a Stuka raid while the Regiment was in leaguer. Hit in the shoulder, he went off proclaiming that he would be back in a fortnight. He had a long journey over rough ground in the ambulance and died that night at Msus.

* * *

The Regiment skirmished on as the Eighth Army developed its

strength to become the outstanding entity of the war, with a 'kind of temperamental climate of its own'. It had its own language, its own uniform (or lack of it) and its own geography. As the Tangier Horse did patrols round Whitehall, York or Whitby, so the Royal Dragoons were familiar with Piccadilly, Knightsbridge and Oxford Circus. They took part in the 'Benghazi Stakes' and the 'Msus Handicap' and sometimes seemed to recreate that headlong dash which had formerly landed British cavalry in trouble but now had its own sufficient reason. British armour was so inferior to German in hitting power that the most our men could do was to dash in for a quick shot or two before the enemy could react.

The see-saw of war in the desert often infuriated the troops and puzzled the public at home, who never really understood the campaign, with its wide swings and the alternate taking and giving of ground. The year 1942 was particularly confused and the Royals, serving when and where and whom they were ordered, often apparently victorious and then ordered to retreat without apparent reason, understood as little of it as anyone. What they did appreciate was an interlude by the sea at Marsa Lucch. The one thing that emerged finally was that General Auchinleck decided to take up the Alamein Line (which was not a line at all) and which his opponent Rommel had decided was his battle line for the conquest of Egypt.

It was perilously like stalemate. Rommel's lines of communication were so stretched that he was fighting on his petrol ration. He could not break through to Cairo, nor could the Eighth Army – or so it seemed – ever get to Tripoli. But the form was changing. Rommel's successes were against opponents poorly equipped, hastily trained and still suffering from a Dunkirk complex – infantrymen bred in the 1914–18 tradition and cavalrymen who still felt the reins between their fingers. The general who converts war into pure science is doomed to failure. It took Montgomery to tip the balance back and over – to make war an art again. He is quoted today as a master of logistics but the operative word is *master*. He made logistics work for him.

Meanwhile the army was being re-equipped with material it could use to effect. The Royal Dragoons got cars they could fight from –

28 Officers of the Regiment visit a well at Shiban in the Hadramut Valley, whilst on a tour of service in the Middle East.

29 Even the most peaceful looking peasants were suspect in the Malay Peninsula. A patrol checks the identity of a passer-by.

30 An armoured car troop of 'C' Squadron on training in Perak State.

31 NCOs of the Royals come home in 1959 to inspect the obelisk in Southwark, on the spot where the Regiment first paraded in 1661.

32 'With Guidon unfurled and swords unsheathed' the Royal Dragoons exercise their right to march through London on 22 October, 1963.

33 'The Last Dragoon' was this trumpeter seen sounding a fan-fare on the submarine HMS Repulse entering Port Canaveral at Cape Kennedy. When he returned to Detmold he found his Regiment amalgamated and the Band to which he belonged dispersed.

Humbers and Daimlers – with a radio set to each vehicle. The vision of Cairo faded: El Alamein commenced to glow as a potential battle honour and the troops would have chuckled had they been allowed to read Adolf Hitler's exhortation: 'There is to be no retreat ... Victory or Death!' Much more to their flippant taste was their general's announcement that they were going to 'hit the enemy for six right out of Africa'. The British are apt to glorify a good fighting opponent if merely to emphasise the fact that they have beaten him. As Napoleon's chief merit is that he met Wellington – once – so Rommel is the man who met Montgomery – more than once. It was the German who went back home.

* * *

There is an unsurpassable description of El Alamein in the Royals' own war history which is too long for reproduction. It opened as a battle strictly for infantry, after a superb overture by the guns. The Royal Dragoons were squatting most uncomfortably between field guns and mediums, with frequent patrols for old-time cavalry reconnaissance. The enemy positions were penetrated but he refused to give way until the bold resolve was formed of sending three armoured car regiments through under cover of darkness. Three commanding officers were summoned to Montgomery's caravan and even he was shaken at Colonel Pepys' plan for his regiment. He proposed to take the Royals clean off the map board – and did just that.

Two squadrons moved off in the wake of the Highland Division ploughing up the powdery sand so that navigation aids were veiled. Twice leading cars fell out of column until at last they came on to a track obviously fit for their purpose. They were almost unmolested, hardly-awakened enemy troops refusing to believe their own eyes or, in one case, taking the Royals for part of their own 90th Light Division mounted in captured British cars. 'Already the east was green behind them', writes the Regimental historian, 'and the waning moon hung like a toy in the sky, but by the time it was light the two squadrons were fleeting through the open desert'. It was real open warfare at last.

The spirit can be recaptured even in such a precise document as a divisional signals' log. Security was of less importance now than getting on with the job, codes being limited to such simple jargon as might puzzle the enemy, while friends would understand. 'The hounds are out of the covert', reported the Royals, 'they turned left and are well away'. An excited liaison officer chipped in: 'The enemy has gone very fast and very far'; while an obvious London Sunday motorist, now Desert Rat, warned: 'We are going to do what the road does at Kingston'.

The Royals fanned out, learning the joys of 'swanning', that Eighth Army word that expressed the antics of venturesome armour in the desert. They enjoyed the superiority of their new cars, their main trouble being from punctures inflicted by enemy air attacks – and even that was at last under some control. They got choosy in the matter of prisoners, taking only good tough specimens, leaving the rest to trudge off on their own – home or to Cairo as they pleased. Rommel himself refused to be convinced – how could he be with people like the Royals slashing every telephone wire? General von Thoma, entertained on calculated grounds at the chief's headquarters, gave rueful confirmation of Montgomery's most resounding victory.

* * *

El Alamein was history: the bells rang out in London; the Army Commander expressed his thanks and the Royals totted up their score. Only the Egyptian businessman was uneasy at the thought that British pockets were no longer available for plundering. *'On craint'*, it was said, *'une victoire premature!'* 'EL ALAMEIN' on the Guidon marks a very definite day's march nearer home. There were still many marches left: while an optimistic allied world deemed it all over bar shouting, the Royal Dragoons joined X Corps for the beginning of a hard campaign, buoyed by the pronouncement: 'We have the chance to put the whole panzer army in the bag'. Weather held them up for days but there was cheering news from Algeria, where new allied forces had landed.

Reading the story of the campaign, it all seems to run so smoothly, relentlessly, the final effort of a superb force. But there were still inevitable checks never to be understood by an exultant nation, realising at last that it had spawned an army of conquering heroes. It did not even look like an army, from the double-badged beret of its commander down to the gay scarves of its subalterns and the illegal grey berets of the Royal Dragoons. It did not worry about what it looked like. All it had to do was to recognise the crawling columns of the enemy. Its lines of communication were less of a problem than Rommel's for, given a coastline along which to fight, the British army looks with confidence to the Royal Navy to see it through. The end run was now certain.

Advancing allied forces were composed to meet the needs of the moment, but the Royals were mainly identified as part of the 4th Light Armoured Brigade, often promoted into a 'force' by getting strange units under command. Their chief made one slip-up in his normally confident predictions. Announcing the forthcoming occupation of Tripoli he proclaimed: 'The KDGs will enter by the east gate, the Royals by the west gate'. But he uttered no complaint when it was the 11th Hussars who entered by the *south* gate. The Royals, indeed, never saw Tripoli, save for a detachment who marched past an exultant Winston Churchill. They were now headed for Tunis, through a country more cultivated, more populous than the desert which had been their haunt for so long.

Two armies met – the First in their neat dress and dark green camouflage and the Eighth from the desert trail. The enemy's plight was now hopeless: he could not even stage a Dunkirk – the Royal Navy hoped he would. Thus, though its number marched on to further victory, the Eighth Army, which had become such a living, pulsing entity, played out its rôle. Three days before a quarter of a million enemy soldiers laid down their arms, 'A' Squadron, the Royal Dragoons left for the already half-forgotten Delta to make ready for the next battle in another continent.

Chapter 14

It was left to 'A' Squadron the Royal Dragoons to represent the armoured cars of the new cavalry in Sicily, with no special preparation save the painting of the cars themselves in 'European' colours. The whole invasion force was drawn from far afield, one formation coming direct from England. Part of the 50th Division, 'A' Squadron embarked somewhat disjointedly from Alexandria, Suez and Haifa and in ten days suffered the pangs of high seas, saw a troop transport torpedoed and were overawed by a purple Etna before they landed. It was a new war fought in largely enclosed country against an army lacking vigorous leadership. Sicily was a prologue to the main Italian act, which opened on 3 September. The Italians were out of the war by the 8th and took up the dubious role of co-belligerents.

The leading squadron of the Royals operated with whatever formation needed their assistance, but rejected the approach of an Italian senior officer who, arriving on a motorbike in boots and spurs, considered them a welcome reinforcement to a private army he was assembling. Memories of desert warfare receded fast but they had no cause to think that their fighting powers would be enhanced by Italian command. They helped to bite off the toe of Italy and drove up the spur. At last, with the approach of winter likely to limit their role and the Italian weather not living up to the blandishments of the travel agencies, they settled down to life on the farm.

The rest of the Regiment languished in North Africa, much cheered by an inspection by the King, their Colonel-in-Chief, who gave permission for the retention of the by now cherished grey beret, until he was reminded from Whitehall that he had himself signed a decree that the only exception to the regulation black of the Royal Armoured Corps was in favour of the 11th Hussars. They were training now from an entirely different book, soldiers enough to know that desert warfare was as much out of date as that practised by their ancestors nearly three centuries ago. They had knocked Hitler out of Africa: they would bury him in the ruins of Berlin if he waited a little. For the massed armies of a modern state a campaign is a painful adventure: for the regular British soldier it is an episode.

The main body of the Royals joined up with their advanced squadron but were spared the arduous fighting up the long leg of Italy and the frustration of the troops left to maintain it. Whatever was said about the 'soft underbelly' of the Axis, the soldiers had the sure feeling that it was to be beaten far to the north and the Royals were among the chosen troops that Montgomery needed around him. Thus, a few days before Christmas they boarded the *SS Volendam* for the Clyde, speeded by the commander-in-chief, who wrote: 'Please tell all ranks how sorry I am to lose them from the Eighth Army. They have always done magnificently and given of their best. Good luck to you all'. He left them to guess the rest.

So home they came. It was a new world, a new Britain, at war in a war which was, to men from Africa, entirely strange. Here were white faces strained by waiting lacking the close fraternity of desert crusaders. There were tea and buns at the dockside, northern hospitality instead of eastern graft, bombs – for those who took London leave – bigger than those they had dodged in real battle. Finally there was training for this new war whose shape few could as yet visualise. They moved south to join XII Corps, took over new, harder-hitting vehicles and were brought up to strength on a new establishment. The grey beret disappeared except for officers, though General Montgomery remarked: 'I shall be pleasantly reminded of old times if the grey beret reappears on the Continent'. But he had to write it down as the battle even he could not win.

* * *

At last, on 27 July, 1944, the Royal Dragoons, with 55 officers and 729 other ranks, arrived off Courseuilles to take part in the greatest military adventure of all time. It could now be reported that 'like a great door hinged on Caen, the whole allied line was swinging forward'. One gem from the regimental account illustrates the almost blind day-to-day give and take in the *bocage*. A Norman peasant who had been employed by the Germans to put up poles to hinder the landing of allied troops, oblivious to the frequent explosions around his home or the change in uniforms, put in his bill to the British for his wages for the month of June.

If ancient battalions were apt to lose their identity in this war of millions, at least armoured cars left scope for individuality – something of the old 'cavalry dash'. With five years of active service to their credit, the Royals could still feel in themselves an afinity with comrades down the centuries. They saw no action until 5 August, when a period opened which was particularly perilous for junior ranks. It was an open question as to whether corporals who, in scout cars were required to lead (as it seemed) a whole army, or sergeants whose normal job it was to search out an enemy flank, had the more dangerous job of it. If they put on weight camping out on Norman farms, they lost it dealing with enemy rearguards fighting from good cover, well-supplied with rockets. Seeing little but the next bend in a narrow road, knowing only the stiffness of continuing opposition, the men in the armoured cars felt none of the excitement only to be gained in the map rooms. When they drove through Falaise it had not become an exclamation mark in history. They saw only the statue of William the Conqueror (that early liberator) that lay among the rubble on the square, his horse's shattered legs pointing forlornly to the sky.

* * *

The advance to the Seine was an exercise in wariness. The enemy's generals were soundly beaten, his soldiers were only capable of carrying on, which they did as desperate individuals. But the Royal Dragoons crossed the great river and led the advance to the Somme. One troop fought its way into the village of La Chaussee Tirancourt where, completely surrounded, it shot its way out with every weapon it possessed, leaving twenty SS dead to mark its passage. And in Picardy the 'cider apples were turning scarlet' and the Royals concealed themselves 'among the Norman cows and dappled Percherons'.

The allies kept up their pressure for open warfare, concentrating on the major task of preventing the enemy from forcing things back to the dreadful stalemate of the 1914–18 trench lines, which the territory they were now crossing brought so much to the mind. The advantage of the armoured car was the opportunity it presented for tip and run tactics: the disadvantage became clear when the opposing infantry realised they had only a lightly-armoured and gunned enemy to contend with. There was a village with a bridge near St Pol which Lt Pelham-Burn investigated. It was full of German troops whom he scattered with light automatic fire and grenades in a swift passage down the street.

His first targets surrendered or ran readily, until a few tougher specimens turned the table and it was the subaltern and his crew that surrendered. The Germans had learned that interrogation of British prisoners was useless. But the stubborn islanders gave themselves away. They were wearing an eagle in their berets with the number '105', so they obviously belonged to the '105 Recce Regiment'. That night their captors released them and gave them a tip of five hundred French francs – thus repaying an unconscious debt from desert days, when the Royals set a batch of inconvenient prisoners loose, gave them rations and water and showed them the way home.

Up they drove into Belgium, where Brussels was newly fallen and the natives called on the Royal Dragoons, for the moment in reserve, to halt and help them celebrate. Their Regimental ancestors had known this country well but they had no time to think of that. There was a great deal of mopping up to do along the coast, which the liberators of the Belgian capital had left untidy. The Royals crossed the Lys firing their seventy-fives over open sights and then, with the eager aid of resistance men, expanded the bridgehead. There were local advances and retirements so that civilian patriots had to be careful which flags they flew from day to day.

On 13 September, 1944, the Royal Dragoons moved to join General Horrocks' XXX Corps, with its 'Wild Boar' sign that had come all the way from Africa. They were to play their part in Operation 'Market-Garden', the most spectacular part of which was the airborne invasion of Holland. At the same time they were rejoined by Major Pepys, who had cajoled authority into overlooking the leg he had lost in Africa and who ended the war as the only armoured commanding officer with that disability.

* * *

The Arnhem story has been told often enough and the Royal Dragoons were not cast for a leading role. The sheer gallantry of the 1st Airborne Division brought them the limelight but perhaps XXX Corps had the harder task and certainly the more sustained one. The two US airborne divisions which had taken the one-road artery up from Belgium were too lightly armed to keep it alone. This was General Horrocks' job, carried out through a land of dykes and ditches: dealing with a now desperate enemy skilfully handled, the threat of the new jet planes operating from their home fields and with worse weather on the way. Once more it became a season of detached engagements where small forces merged, fought and dispersed at need.

Lt Goodall, for instance, had a minor war in the village of Eerde one day, being driven out with the loss of both his armoured cars. A mile back his scout car was commanded by Major Balfour. This was a corporal's job but this newly-appointed squadron leader had chosen this way of learning his job. He scraped up sixty willing anti-aircraft gunners, a mixed bag of twenty-five semi-combatants and some Americans who included a general's ADC. Three tanks thickened the defence as German infantry, backed by guns, made their assault. Throwing the enemy back, Balfour put in a three-pronged counter-attack and reopened the road. Goodall and his operator were taken prisoner but escaped in the confusion.

Arnhem deteriorated into near-disaster and over the scattered disputed territory the Royal Dragoons had an unenviable time of it, their armour often little more than a means of transport from one danger area to another. The allied forces were more dangerously stretched than is generally recognised and Twenty-One Army Group made the pleasant discovery that an armoured car regiment could, at a pinch, hold twelve miles of front. In addition such a regiment, with its efficient radio net and young officers and NCOs with more experience than was general, could be called upon to take command of a mixed force assembled for any purpose.

There were incidents enough to liven the story; Eindhoven was a haven where there were cinemas, canteens and girl friends: one corporal set a new fashion by requesting permission to marry but had his eagerness set back by a sudden order to move. An officer who had set fire to a cottage with tracer ammunition took advantage of the calamity to hang up his sodden clothing to dry, whilst one troop made a dart into Germany and put up in the house of an elderly aunt of Hermann Goering. She assured them they would be quite safe since the Reichsmarschall had given instructions that the house was not to be shelled. Civil life still persisted in the war zone to the extent that the verger of Hilvaranbeek complained that near-by gunners were doing their ranging at the very hour reserved for choir practice.

Nothing could illustrate the ups and downs in the life of a regiment better than the dreary Dutch winter. Men who had started the war on horseback in Palestine now wore snow-camouflage suits to trudge over a Christmas card landscape where the thermometer sometimes registered twenty-five degrees of frost. Royal Dragoons lined the banks of the Maas, dashed into northern Holland and learned the hard way that there was still kick enough in the *Wehrmacht* to make life unpleasant. Even their New Year was spoiled by a reckless firework display from the enemy and an amplified broadcast from a morose *Führer*.

* * *

But the end, though they could be excused for not seeing it, really was in sight. Even if the British Rhine crossing had some of its thunder stolen by the American success at Remagen, it was the reward for all the desperate efforts and still more desperate hopes of a people and its forces that had known no let-up since September, 1939. For the Royals it may have seemed unspectacular, since their main job was marshalling the army over the Rhine at Rees, with two thousand men under Regimental command. But it was a job that only one of the best regular regiments could do adequately.

Then on into the fat lands of Westphalia and Hanover, over which their ancestors had ridden and 'where the trees were bursting into leaf and fat cows munched the pasture and small children, following the usual custom, waved at the soldiers'. There was some return to civilised warfare, as when Major Fisher, having seen a sergeant of the Regiment wounded in an ambush, sent over a white flag to get him back. He came in a German ambulance, whose driver was given breakfast and a tank full of petrol. There was still fighting enough but the enemy was seeing sense at last and on 2 May, 1945, the bag of the Royal Dragoons was ten thousand prisoners. Then, just one day before the end of it all, another move was laid on.

Once more there was a new country, a small, welcoming patch of earth to be returned to its people. On 7 May, 1945, the Royal Dragoons crossed the frontier into Denmark as representatives of the British Army. This time there was no anti-climax. Tunis had not cared very much who won the war in North Africa: the Danes did – and showed it. 'No victorious army', wrote the Regimental historian, 'can ever have received such a friendly, spontaneous or wholehearted welcome as that which the Royal Dragoons found in Denmark'. The fruits of peace were sweet but melted too rapidly: by early November they were back in Germany. It was the usual round in an army that must rebuild itself and do a responsible job at the same time. Europe had to be cleaned up while regiments almost disintegrated in the throes of demobilisation and there was the ever-present query – what will the Russians do next? It has never yet been answered. For the Royals it constituted a five-year stretch of work.

* * *

Chester, their first home station (in November of 1950) looked good to them and their stay was high-lighted by a visit from King George VI: but the great let-down was near. Three months at home – and they were off to the Canal Zone, no longer the gateway to the 'Shiny East' but part of a hostile land growing daily more unfriendly, culminating in the Egyptian abrogation of the treaty covering the presence of British troops. Terrorism never ceased but eased down after a British force, including a detachment of the Royals, flushed out the notorious Police Caracol at Ismailia. Finally, in the same *Empire Ken* which had brought them out, the Royal Dragoons left for home on 28 January, 1954, gazing out curiously as they sailed by Tangier, their first foreign station.

It was at Tidworth, on 27 April, that they received their new Guidon from Field Marshal Sir John Harding, with its brave new blazons telling the story of yet another honourable achievement. The Guidon was brought on parade by the Quartermaster Major C. W. J. Lewis, the only member of the Regiment on parade who was serving in 1925 when the old Guidon was presented by King George V. The Royal Dragoons were allowed just three months at home and were then sent back to Germany – so familiar to British troops that it was hardly accounted 'abroad'.

Cavalry regiments were now given at least a 'home' address, which served as a centre for recruiting and linked them up with what remained of the Yeomanry. The Royal Dragoons, the only armoured cavalry regiment selected for Eastern Command, were first based at Bromley, Kent with a recruiting area covering Surrey, Sussex and Kent and including their home town of Southwark. They were affiliated with the Kent and County of London Yeomanry (Sharpshooters) and on the latter being reduced to squadron strength in 1965, moved with them to Croydon. But they were unluckier than some other regiments in that it was not possible to find a home for the trophies which they had accumulated.

The next tour of foreign service took them in the old familiar direction – East – first to Aden, with one squadron detached to Sharjah in the Persian Gulf. Still seeking the 'Shiny East' they went on, in October, 1960, to Malaysia, with a squadron in Singapore, staying for two years. It was just another spot of trouble such as most regiments were familiar with – winkling terrorists out of their lairs, patrolling the everlasting jungle of the long peninsula, hardly finding time to remember that they had a three hundredth birthday. Home they came again in 1962 – it was Tidworth again, that station on the Plain which only determined soldiers could make habitable. It was here they received their first tanks, only to have two squadrons hastily mounted again in armoured cars for a spell of duty in Cyprus.

London had not forgotten its Regiment and had bestowed its Freedom on the Royals in 1961 in honour of the Regimental tercentenary. It was an honour not lightly given, for the City scrutinised the antecedents of the Royal Dragoons back to their first muster in Southwark. Now, with the Regiment finally assembled at Tidworth, the Royals went all out to show the Lord Mayor and their fellow citizens that pride was two sided. They drilled, rehearsed and were tailored by experts. Officers brushed up on sword drill and extra drummers were trained to supplement the normal meagre allowance of a dismounted cavalry band. For the terms of the Freedom were that they were privileged to march through Town with 'Drums beating, Swords unsheathed and Guidon unfurled'. They even borrowed a drill sergeant from the Grenadier Guards, determined that, as another drill instructor of another dismounted cavalry regiment had once said: 'If we *must* be infantry then we'll be the *best* infantry'.

Thus, on 22 October, 1963, the Royal Dragoons came home, to march past their Lord Mayor on his own doorstep. 'As we swung through the gate (of Armoury House)', an officer wrote, 'all London burst in on us and carried us along'. They were challenged at Ropemaker Street by the City Marshal and then, drummers throbbing that indescribable marching rhythm of the British Army, the Band giving its every breath to the Regimental March, the Royal Dragoons gave their 'Eyes Right' to the Lord Mayor and were Freemen of London. It had been a long road through three centuries of history and the Royals had upheld their pride in more than thirty countries.

Pride burst out in their message to Her Majesty the Queen.

'The Royal Dragoons exercising their Right of entering the City for the first time send humble and Loyal Greetings to Your Majesty'.

Epilogue

'Whatever changes may come through evolution in the use of cavalry in the future, the Royals should be the last regiment to be disbanded, as they were the first to be raised and in the years to come a future historian may yet be able to add chapters which will continue the glorious record of the past'.

So wrote, in 1933, Brigadier Ernest Makins, then Colonel of the Royal Dragoons. He need, as will have been seen in this short account, have had no fears as to the chapters to be added but his words show the uncertainty in which the cavalry of the 'tween-war period lived and trained. They were shaken by the amalgamations of 1922 and dreaded what might come. Disbandment has never threatened the Royal Dragoons seriously and they were almost the last to be amalgamated.

Vesting Day for the new Regiment of the Blues and Royals was 31 March, 1969, amalgamation having taken place officially the day before. Then, on a barrack square in Detmold the Regiment paraded in its new guise and, after the reading of Her Majesty's message the Colonel of the amalgamated Regiment, Field Marshal Sir Gerald Templer addressed the parade: 'Gentlemen of the Blues and Royals: I don't think a speech is really necessary because we all know what it is about – and what we're at but there are one or two things I must say . . . Up to the day before yesterday we were two proud and individual regiments, each with three centuries of loyal and distinguished service to our Sovereign and our country.

'We now enter on another phase of our being – as The Blues and Royals (Royal Horse Guards and 1st Dragoons). The late Blues are immensely proud to wear the Eagle and the Dettingen black patch of the Royals and I know the late Royals are equally proud to wear the embellishments of The Blues. It is indeed a privilege for us all to form part of the Household Troops and that honour carries with it corresponding responsibilities, as you will all realise . . .

'For me it is a great honour to have been appointed your Colonel and it is an especial satisfaction to me that, with the approval of Her Majesty, General Sir Desmond Fitzpatrick (last Colonel of the independent Royal Dragoons) is the Deputy Colonel. I have not the slightest doubt that our new Regiment will not only maintain the traditions of the past but will set the highest standard of loyalty and military efficiency in the future.

'Gentlemen of The Blues and Royals – it is up to us to achieve that and I have no doubt that you, the serving members, will ensure it'.

* * *

Southwark, 21 October, 1661 – Detmold, Germany, 29 March 1969 – these are the datelines of the Regiment's independent career. Tangier Horse, Royal Regiment of Dragoons, now they are back as Horse again, still 'Royal' as a component of the Household Cavalry. They have changed the scarlet tunic for the blue but still wear on their sleeves the proud trophy they won at Waterloo, the Eagle.

Regimental Music

British regiments have always been left very much to their own devices in the matter of providing music, at least until the late Duke of Cambridge, incensed by the performance of regimental bands at a review during the Crimea War, took a hand in establishing Kneller Hall as a centre for regulating it. Dragoons were however helped by the fact that they were expected to march on foot as well as mounted and were allowed the generous establishment of one drummer and one 'hautboy' per troop – a total of twelve nominal musicians per regiment. It was nevertheless a distinct grievance to the Royal Dragoons that, on ceasing to be the Tangier Horse they lost the coveted privilege of marching behind kettledrums. Defying regulations they provided themselves with drum horse and kettledrummer as early as 1704 in Spain and were soon boasting of their 'Musick'.

In 1766 dragoon regiments were allowed to substitute trumpeters for drummers and hautboys and regimental accounts for the 1770s record the purchase of a 'fife, bassoon, clarinet and horns' whose players are reported as being 'expensively dressed', naturally at the expense of the officers, or at any rate of the colonel. A year or two later the trumpeters are reported as being 'handsomely dressed and form a good band of music'. The libel was still being perpetuated that no Englishman could really understand (or at any rate play) music and consequently all the trumpeters of the regiment were Germans and it is possible that when, in 1802, the Regiment was 'allowed to enlist a person for five years to instruct the Band' he was a German too.

Trumpeters were for long mounted on grey horses, as they are today in the Household Cavalry but the great pride of cavalry regiments was always the Drum Horse. 'Danny' who served in the Royals up to 1889 and 'Jack', a piebald of later on in the century have come down in regimental memory but the most famous was 'Coronet' presented by King George V and one of the last of the Hanoverian creams, who joined from the 2nd Life Guards in 1922 and 'soldiered on' until he died in harness abroad.

An ex-trumpeter of the Royal Dragoons claims the honour of being 'the Last Dragoon'. With amalgamation imminent he received orders to report for a 'flag-showing' voyage of the Royal Navy Submarine *Repulse* to America. When the vessel came into Cape Canaveral Submarine Base it was a Royal Dragoon in full uniform who saluted the United States of America from 'forward' and then returned to Detmold to find amalgamation complete 'with no Band to return to and a lot of new faces'.

The Royal Dragoons are rich in regimental music and may possess the oldest British cavalry march in 'Dragoons of Tangier' composed in 1696 and arranged for modern instruments by Bandmaster Mackay. The Regiment at one time used Gounod's 'Soldiers' Chorus' as their march and there are two more – 'Spectemur Agendo' and 'The Royals' – both by Trythall. Finally, as both slow and quick marches Blankenburg's 'The Royal Dragoons' was composed during the colonelcy-in-chief of Kaiser Wilhelm II and is now used, in conjunction with the marches of the Royal Horse Guards, in the amalgamated Regiment of The Blues and Royals. The arrangement which follows is that of the slow march.

Regimental Slow March

Blackenburg's 'The Royal Dragoons'

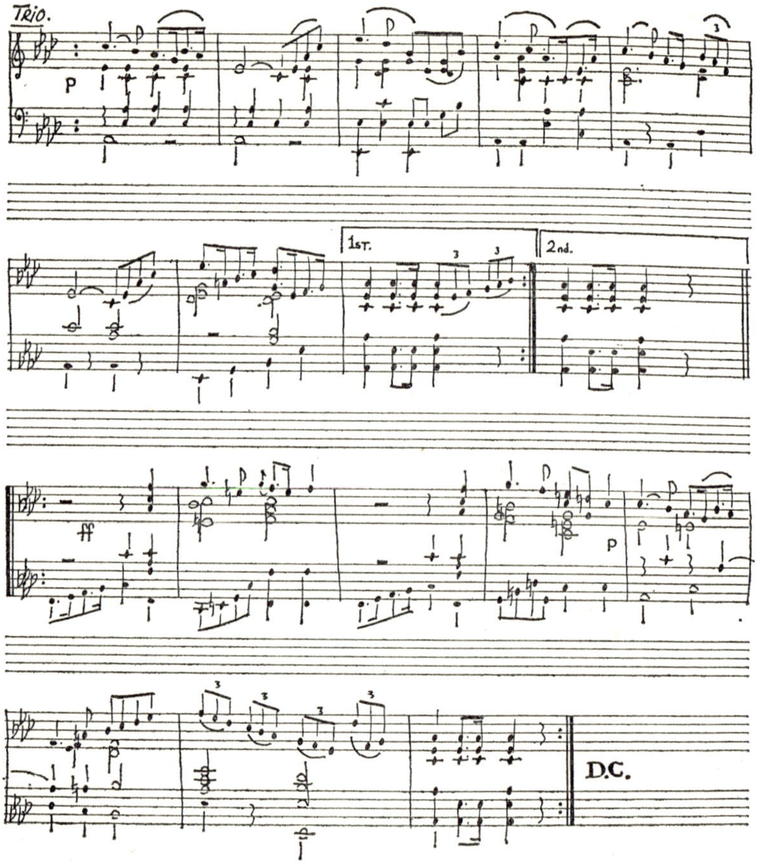
TRIO.
p
1st.
2nd.
ff
p
D.C.

Battle Honours

'TANGIER, 1662–1680'; 'DETTINGEN', 'WARBURG', 'BEAUMONT', 'WILLEMS', 'FUENTES D'ONOR', 'PENINSULA', 'WATERLOO', 'BALAKLAVA', 'SEVASTOPOL'. 'RELIEF OF LADYSMITH', 'SOUTH AFRICA, 1899–1902'.

First World War:
'YPRES, 1914, '15', Langemarck, 1914', 'Gheluvelt', 'Nonne Bosschen', 'FREZENBERG', 'LOOS', 'ARRAS, 1917', 'Scarpe, 1917', 'SOMME, 1918', 'St-Quentin', 'Avre', 'AMIENS', 'HINDENBURG LINE', 'Beaurevoir', 'CAMBRAI, 1918', 'PURSUIT TO MONS', 'FRANCE & FLANDERS, 1914–18'.

Second World War:
'NEDERRIJN'. Veghel', 'RHINE'. 'NORTH-WEST EUROPE 1944–45', 'SYRIA, 1941', 'Msus'. 'Gazala', 'KNIGHTSBRIDGE', 'Defence of Alamein Line', 'EL ALAMEIN', 'El Agheila', 'ADVANCE ON TRIPOLI', 'NORTH AFRICA, 1941–43', 'SICILY, 1943', 'ITALY, 1943'.

(*Note*: Of the Honours granted for the two world wars, only those printed in capitals are borne on the Guidon).